The ArcGIS® Book

10 Big Ideas about Applying Geography to Your World

Esri Press
Redlands, California

All images courtesy of Esri except as noted

On the cover:
The Stamen Design Watercolor Map applies raster effect area washes and organic edges over a paper texture to create an effect reminiscent of hand-drawn maps. Created with OpenStreetMap data and published as a map service for use as an alternative basemap in ArcGIS Online.

Table of Contents

Introduction

See this book come alive at www.TheArcGISBook.com

You are reading the print version of this book. It exists wholly online at TheArcGISBook.com. Bookmark it now for when you're ready to sit down at your computer and work with ArcGIS.

Big things are happening in the world of maps and mapmaking. A convergence of technology and social trends has pushed geographic information systems (GIS) onto the Internet in a significant way, and the vision of a global Web GIS has been realized. This book explores ten "big ideas" that encapsulate these trends, and shows you how to apply these ideas to your own work, in your own world.

The Web GIS revolution is radically altering how information about the world around us is applied and shared.

This is a book about ArcGIS, the Web GIS platform. But ArcGIS is more than just mapping software running online. It's actually a complete system for discovering, consuming, creating, and sharing geographic data, maps, and apps designed to fulfill particular objectives.

The twin goals of this book are to open your eyes to what is now possible with Web GIS, and then spur you into action by putting the technology and deep data resources in your hands via the Quickstarts and Learn ArcGIS lessons that are included in each chapter. By the end, if you complete all of the exercises, you'll be able to say you published web maps, used story maps, built a 3D cityscape of Venice, Italy, configured a custom web app, performed sophisticated spatial analysis, and much more.

Once the exclusive realm of technologists, digital mapmaking has gone mainstream, empowering everyone.

The basics of ArcGIS are easy, engaging, and fun, and even more sophisticated features (like spatial analysis and web app development) are now accessible to everyone, not just the experts. With the world's geography at your fingertips, you'll be empowered to affect positive change in the world around you.

Freely available and approachable, Web GIS makes for a kind of democratization of mapping and analysis of the world around us. If we think of geography as the ultimate organizing principle for the planet, then Web GIS is the operating system. The challenges we face, from our local neighborhoods to our world as a whole, all share the commons of geography: they are happening somewhere, which places them squarely "on the map."

How this book works

Who is the audience?

This book has been designed with several audiences in mind. The first is the professional mapping community—the people who create or work with geospatial data as a dedicated activity—in particular those GIS professionals who are just beginning to leverage online mapping. The second is the broader world of web technologists, information workers, web designers, and Internet-savvy professionals in every related field. The technology has become so ubiquitous and easy to use, a third audience is really any individual with an interest in maps and an idea for how to apply it. The only prerequisite is a desire to better understand online mapping and a roll-up-your-sleeves attitude.

Learn by doing

This is a book that you do as well as read, and all you really need is a personal computer with web access. The adventure starts when you engage yourself in the process by doing the lessons in this book. Each step of the way you will gain new skills that take you further. Mapping professionals are in high demand for a reason. Businesses, governments, and organizations of all stripes can see the utility. This book is a call to action and a blueprint for how to get there. It's about applying geography to your specific situation, problem, or conundrum, and finding a solution with Web GIS.

While reading this on one of several available platforms, including print, you can practice making maps on the web with your computer. With the Interactive Edition of the print book on the Web, you will experience and use many of the example maps and apps as they come to life on the screen.

In each chapter, the Quickstarts tell you what you need to know about the software, data, and web resources that pertain to that aspect of the ArcGIS system. The Learn ArcGIS Lessons pages are your gateways to online instructional content from the Learn ArcGIS website.

While structured with one big idea per chapter, each chapter provides many more granular ideas. Open the book and read any page, or read it front-to-back and be part of the adventure every step of the way. Experience Web GIS at your own pace according to your own interests.

More than anything else, we want you to feel empowered to dive right into ArcGIS and expand your horizons by doing real mapping and analysis with Web GIS. What problem in your life or within your purview would you like to scope out? If it has a geographic element (and most do), then it's something you can tackle with GIS.

A word on devices and capabilities

All of the web-based functions on ArcGIS Online are accessible through standard web browsers on Mac, Windows, and Linux devices. The apps run on iOS and Android computers. Desktop applications ArcGIS 10.3 and ArcGIS Pro are for PC Windows machines.

Maps, the Web, and You
Power and possibility with Web GIS

The phenomenal growth of consumer web mapping has opened the eyes of the world to the value of maps and geography and created an audience ready for more sophisticated spatial analysis and geographically oriented storytelling. That's where you come in.

Geography applied
Web GIS is available to everyone

This first chapter shows you, a current or aspiring GIS user, how the new Web GIS paradigm enables you to contribute your local wisdom to the whole. If you are an experienced hand, you may think of GIS as a series of software programs and tools that you use to get mapping and data editing work done. This is often accomplished in workgroup or enterprise settings using traditional, back-office computing environments. But the technology has evolved, and while many of the traditional workflows still have a role, there's also now an expanded vision for how the world works with your geographic information, in concert with everyone else's geographic layers, using ArcGIS, an online, cloud-based platform.

ArcGIS provides an open computing platform for maps and geographic information, making it easy to create and share your work as useful, interactive GIS maps, data layers, and analytics. The big breakthrough is in how easy it is to engage with the system. This ease of use is not just about improvements in the software user experience and interface, but about how the system is directly connected to and interoperable with an authoritative and collective geographic data repository of immense proportions.

All of the advanced geographic intelligence that you create as data, maps, and analytical models can be mashed up and delivered as online maps and apps and shared with others who can put them to work—both within your organization and beyond. The people who need your information can work with easily configured apps on their computers, tablets, and smartphones to bring your geographic information to life.

Whether you are that experienced power user, who remembers ArcInfo and cut your teeth on the ARC Macro Language, or a newbie about to make your first map, this book is your introduction to a compelling new vision for how the world works with geographic information. We refer to this as Web GIS.

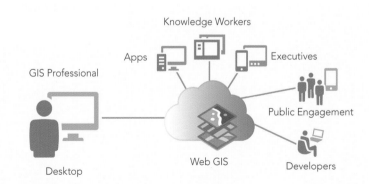

The GIS professional working on the desktop creates and shares information to the Web GIS (which can be in the cloud, on your secure local network, or both). Knowledge workers, executives, citizens, developers, and other GIS users can build upon and leverage your work. In turn, you can leverage other users' layers for your own work as well.

Thought Leader: Jack Dangermond

Web GIS is a new pattern for applying geography

In the past few years, the considerable growth in consumer web mapping has not only opened the eyes of the world to the value of maps and geography, it has likewise created a new pattern for the GIS community to easily disseminate and make available its knowledge as high-quality maps and GIS analytic services. This pattern is improving how we communicate our world. It will also allow neogeographers to integrate and leverage the work of the traditional GIS community.

At Esri, we've designed our business around thinking first about the people who use our GIS tools. As we've grown from hundreds of users to thousands and now millions, we've continually invested in research and development. The result of all this attention and investment is a Web GIS platform that we feel is unrivaled in its capabilities and potential. And we've actually only just begun this move onto the web. In the coming years we aim to keep innovating to bring the full strength of all our traditional desktop tools to the web. It's an exciting thing to be a part of.

I feel very strongly that our worldwide Esri user community—which hopefully includes you—is uniquely positioned to face the serious challenges ahead. My hope is that your authoritative data and thoughtful analysis of whatever it is you study, track, and monitor will have a positive impact on the future.

Jack Dangermond is president and founder of Esri, the world leader in GIS software development and its application in business, health, education, conservation, utilities, military and defense, oceanography, hydrology, and many other fields.

 Watch a talk by Jack Dangermond
Esri.com/ArcGISBook/Chapter1_Video

Web GIS is collaborative
Geography is the key, the web is the platform

Every day, millions of GIS users worldwide compile and build geographic data layers about topics critical to their work and for their particular areas of interest. The scope of information covers almost everything—rooms in a building, parcels of land, neighborhoods, local communities, regions, states, nations, and the planet as a whole. Web GIS operates at all scales, from the micro to the macro.

Geography is the organizing key; information in Web GIS is sorted by location. Because all these layers share this common key, any theme of data can be overlaid and analyzed in relation to all other layers that share the same geographic space.

This is a powerful notion that was well understood by mapmakers in the pre-digital era: tracing paper and later Mylar and other plastic sheets were employed to painstakingly create "layer sandwiches" that could be visually analyzed. The desire to streamline this process using computers led to the early development of GIS. The practical term for this notion is "geo-referencing," which means to associate something with locations in physical space.

Now extend the idea—of geo-referencing shared data—onto the web. Suddenly it's not just your own layers or the layers of your colleagues that are available to you, it's everything that anybody has ever published and shared about any particular geographic area. This is what makes Web GIS such an interesting and useful technology; you can integrate any of these different datasets from different data creators into your own view of the world.

All GIS data fits onto the Earth's surface

Everyone's data can be integrated and used

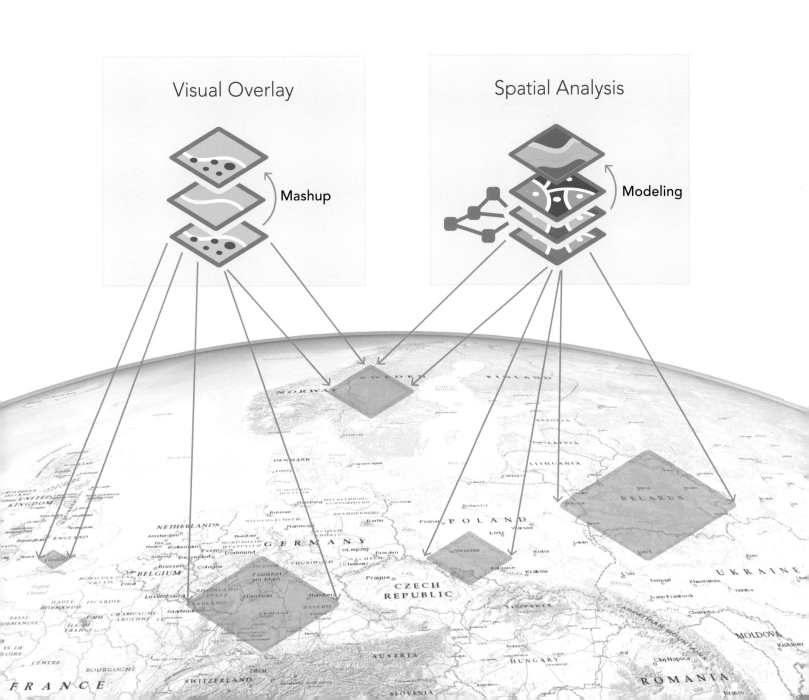

The expansive reach of Web GIS
Across organizations and beyond

The geographic organizing aspect of GIS has been part of GIS thinking from the beginning, but now factor in the impact of the web. The new Web GIS provides an online infrastructure for making maps and geographic information available throughout an organization, across a community, and openly on the web. This new vision for Web GIS fully complements, integrates, and extends the work of existing GIS professionals.

Web access to data layers is straightforward: every layer has a web address (a URL) making it easy to locate and share online. And since every layer is geo-referenced, Web GIS becomes an engine of integration that facilitates the access and recombination of layers from multiple providers into your own apps.

This is significant. Millions of professionals in the GIS community worldwide are building layers that serve their individual purposes. By simply building and then sharing these layers back into the GIS ecosystem, they are adding to a comprehensive and growing GIS of the world. Each day, this resource grows richer and is tapped by ArcGIS users and shared on the web. Web GIS truly has become "the nervous system of the planet."

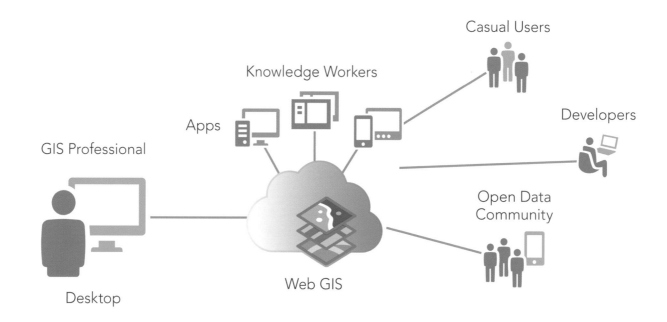

Web GIS extends the reach of the work of GIS professionals to others inside of their organization and to their constituents and beyond.

GIS is evolving
The new ArcGIS is a Web GIS

GIS has already evolved and continues to do so. Its information model was originally centered around local files on a single computer. GIS evolved from there into a central database environment that revolved around clients and servers. The most recent evolutionary stride has taken it to a system of distributed web services that are accessible in the cloud.

ArcGIS has become a Web GIS platform that you can use to deliver your authoritative maps, geographic information layers, and analytics to wider audiences. You do this by using lightweight clients and custom applications on the web and on smart devices, as well as desktops, as you'll see in later chapters.

Much of the core work of traditional GIS users and experts has involved building and maintaining key foundational layers and basemaps that support a particular agency's operations. Billions have been invested in compiling these basemaps and data layers in great detail and at many scales. These include utilities and pipeline networks, parcel (land ownership) fabrics, land use, satellite imagery and aerial photography, soils, land cover, terrain, administrative and census boundaries, buildings and facilities, habitats, hydrography, and many more essential data layers.

Increasingly, these layers are finding their way online as maps, comprehensive data layers, and interesting analytical models. This data comes to life for everyone as a living atlas, a collection of beautiful basemaps, imagery, and enabling geographic information, all of which are built into the ArcGIS platform. There, they are available for anyone to use, along with thousands of datasets and map services that have also been shared and registered in ArcGIS by users like you from around the world.

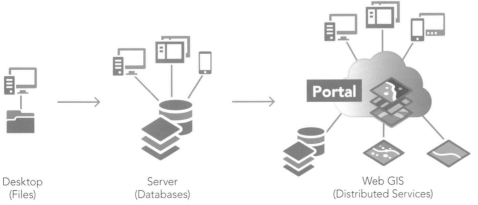

Desktop
(Files)

Server
(Databases)

Portal

Web GIS
(Distributed Services)

Geospatial information has evolved from files to databases and now to the web. Each organization's individual GIS becomes part of a synchronized global platform. Data layers and analytics created by individual organizations are arranged into themed collections on the web. Every item has a URL. There is a data catalog, a searchable portal where you can discover data created by others and combine it with your own data in a host of GIS applications.

ArcGIS information items
Maps, scenes, layers, analysis, and apps

Think of the information items eligible to be stored in ArcGIS as different types of geographic information. Let's examine some of the most important ones: web maps and scenes, layers, analysis, and apps. Go to www.TheArcGISBook.com to see them come alive.

Web maps and scenes

GIS maps and scenes (their 3D counterparts) are the primary user interfaces by which work is done with ArcGIS. They contain the payload for GIS applications and are the key delivery mechanisms used to share geographically referenced information on the ArcGIS platform.

Every GIS map contains a basemap (the canvas) plus the set of data layers you want to work with. If it's 2D, it's called a "web map." These are examples of two-dimensional web maps.

US Population Change

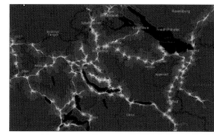

Highway Access in Europe

US Minority Populations

If it's in 3D, it's called a "scene." Scenes are similar to web maps (they combine basemap layers with your operational overlays), but scenes bring in the third dimension, the z-axis, which provides additional insight to study certain phenomena. These are examples of scenes.

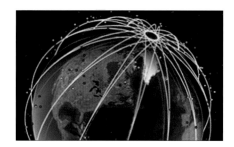

Satellite Map

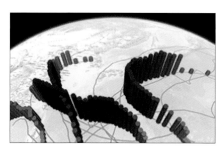

Pacific Typhoons

Election Results in 3D

Layers

Layers are logical collections of geographic data. Think about any map. It might contain such layers as streets, places of interest, parks, water bodies, or terrain. Layers are how geographic data is organized and combined to create maps and scenes; layers are also the basis for geographic analysis.

There are many types of layers. They can represent geographic features (points, lines, and polygons), imagery, surface elevation, cell-based grids, or virtually any data feed that has location (weather, gauges, traffic conditions, security cameras, tweets, etc.). Here are a few layer examples.

Nepal earthquake epicenters

Feature point data from in-ground data sensors.

Toronto traffic

Cell-based raster using historical predictive data.

Terrain of Swiss Alps

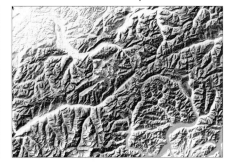

Tinted hillshade is a cell-based raster derived from elevation surface.

Stanford University buildings

3D forms procedurally generated using Esri CityEngine rules.

Sioux Falls parcels

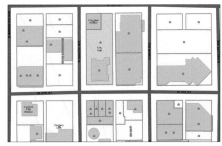

Feature polygon data from cadastral surveys of Sioux Falls, South Dakota.

New South Wales wildfire tweets

Feature point layer of tweets during 2013 New South Wales fires at #SydneyFires.

Analysis

GIS analysis is the process of modeling spatially, deriving results by computer processing, then examining and interpreting those model results. Spatial analysis is useful for evaluating suitability and capability, estimating and predicting, interpreting and understanding, and much more.

ArcGIS includes a large set of modeling functions that produce analytical results. These typically generate new data layers and associated tabular information in your GIS, enabling you to use ArcGIS to model just about any kind of spatial problem you can think of. (Chapter 5 delves into this area of ArcGIS in more detail.)

Sometimes analysis functions are built into the system like the workhorse Create Buffers tool in ArcGIS Online. In many other situations, experienced users create their own models as analysis tools that can be shared as geoprocessing packages with other ArcGIS users. These can also be used to create new geoprocessing tasks in ArcGIS Server. In other words, advanced users can create sophisticated analytical models that can be shared and accessed by other users who can work with their results.

This means that even beginners can apply spatial analysis. Practice and experience will help you to grow the level of sophistication of your spatial modeling. The cool news is that you can begin applying spatial analysis right away. The ultimate goal is to learn how to solve problems spatially using GIS.

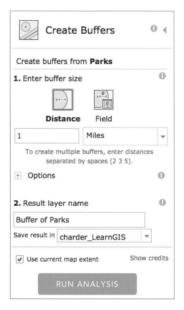

Commonly used tools like Create Buffers are built into the software.

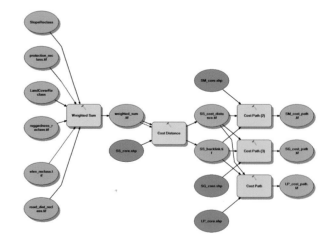

Specialized custom tools created by users are shared as geoprocessing packages online. They can also be shared as geoprocessing tasks in your portal.

Apps

ArcGIS apps are lightweight map-centric computer programs, designed to run on smartphones, tablets, and other mobile devices. We'll talk about apps in depth in chapter 7, but now you should know that as a publisher in ArcGIS, you can configure an app for specific users you want to reach by including a certain map or scene, data layers, and setting other app properties. These configured apps are what you can save and share with selected users. And you can manage these as app items in your ArcGIS portal.

One consequence of the explosion of smartphones and devices has been the "app revolution." Every consumer can find and use focused, single-purpose apps that enrich their daily lives. Interestingly, mapping apps have been among the most popular apps. Everyone knows how to use maps in their daily lives.

So, many already love and appreciate working with these map-based apps, and want map apps that "do more." There's a growing awareness of the value that comes from extending the use of mapping tools into work and organizational settings.

It's not surprising to realize that most GIS apps are also based on the use of maps. If a map has a purpose and an audience—and a user experience—it's an app. And just like consumer-based mapping apps, these are familiar and easy for people to adapt and use in their work lives as well.

The result is that map-based apps are the way that GIS organizations are extending the reach of their GIS in significant ways.

How ArcGIS organizes your content

Portal provides galleries for organizing and sharing GIS content

GIS is collaborative and social. Everyone depends on other users for some of the geographic information they need to do their jobs. Beyond your own personal information items, how do you find and discover these other sources? They exist both in your organization and on the web, but what is the mechanism that makes this available to you and your work?

The portal is your GIS information catalog and helps you by organizing access to information layers and by making your maps and geographic information items available to others. Portal enables collaboration.

Portal organizes content into galleries as information items of various kinds. Through these galleries, you have access to your own personal content items ("My Content"), to your organization's items, and to items that are shared by the broad ArcGIS community. All of these information sources are critical to your work.

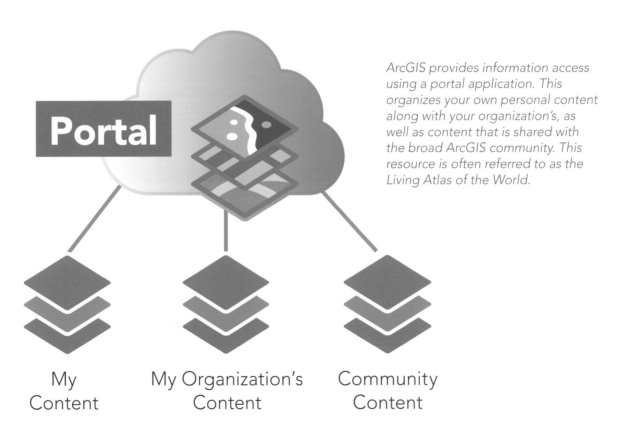

ArcGIS provides information access using a portal application. This organizes your own personal content along with your organization's, as well as content that is shared with the broad ArcGIS community. This resource is often referred to as the Living Atlas of the World.

My
Content

My Organization's
Content

Community
Content

It all begins with a map

Web GIS revolves around the map. It's framework for your data and the primary geographic container that gets shared and embedded in your apps. In ArcGIS it is called a "web map." The purpose of the web map below is simple enough: to show the last 60 days of earthquakes everywhere on earth. (By the way, if you're reading the print edition, make sure you access the book on your computer as well to get the full experience.)

There are several points of interest right here on this map. First of all, it's navigable, which means you can pan and zoom. The map actually has many zoom levels, each level revealing more detail the closer you get. Click on any earthquake symbol to learn the magnitude and date of each event. These little windows of information are called "pop-ups," and by the time you finish this chapter, you'll know how to configure them.

The map also has scaled symbols, showing the relative magnitude of each earthquake. The background map is symbolized as well, in this case in muted dark tones that set off the bright earthquake symbols.

This data, organized with this combination of symbology, reveals an interesting pattern: the well-known Ring of Fire. This map could be easily embedded on any web page. But where did it originate? It began life as a web map in ArcGIS Online, in the Map Viewer.

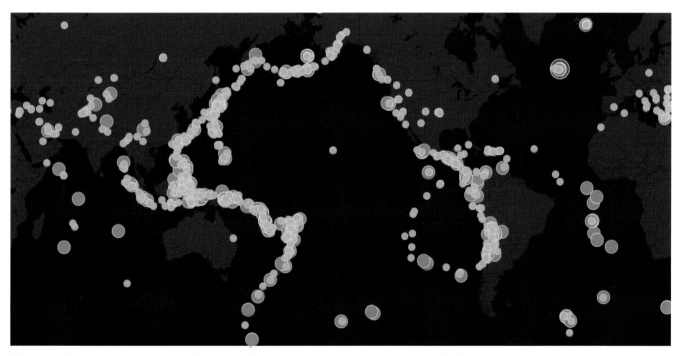

This simple live map shows earthquakes that happened over the previous 60 days.

Quickstart

Connect with and deploy the ArcGIS platform

Now it's time get your hands on ArcGIS. If you're an existing user and already have an ArcGIS subscription (with Publisher privileges), as well as the ArcMap and ArcGIS Pro desktop apps installed on your local machine, you're good to go and can skip to the next page. If you don't have these three things, read on.

▸ **Get a Learn ArcGIS organization membership**
The majority of lessons in this book are carried out on the ArcGIS platform (in the cloud), and require membership (with Publisher privileges) in an ArcGIS organization. The Learn ArcGIS organization is available for students and others just getting started with ArcGIS. With your membership, you can immediately begin to use maps, explore data resources, and publish geographic information to the web. Go to the Learn ArcGIS organization and click the Sign up now link to activate a 60-day membership.

Getting a Learn account is the quickest and easiest way to experience web GIS at ArcGIS Online. To get the desktop applications ArcMap and ArcGIS Pro (used in the lessons for chapter 2, 7, 8, and 9), you'll need to activate the ArcGIS Trial.

▸ **Activate the ArcGIS Trial**
If you think you want to eventually get your own ArcGIS organization (for example because you know you'll be the administrator), or if you want to do ArcGIS and Pro projects, then instead of using the Learn ArcGIS organization, you'll need to activate an ArcGIS Trial. The trial activation will create an ArcGIS organization (with your very own URL that you choose) that will be your own personal sandbox (for 60 days). If you want more information about doing this, refer to page 139.

The advantage is that if you decide you'd like to continue, you can purchase an ArcGIS account and make your web site permanent, thus perserving any web maps, data layers, and apps that you create.

▸ **Using a Public Account**
An alternative way for students, individuals, and educators who prefer not to start a 60-day trial subscription is to sign up for a Public Account. You won't be able to do all of the lessons in the book, but there is still a lot to see and do.

Learn ArcGIS Lessons

Map and analyze lava flow hazards on the Big Island

In these two projects, you'll get to spend some time on the Big Island of Hawaii, learning how to make maps in an ArcGIS organization. In Get Started with ArcGIS Online, the first project, you will map lava flow risk in an actively volcanic part of the world. In the second project, you will analyze emergency shelters.

▸ **Overview**

You'll get acquainted with the island's volcanoes and geology as you explore a map, make your own map, and work with its symbols and pop-ups. You'll turn table data from a Comma Separated Values (CSV) file into spatial information, and you'll package a web map as a professional-looking application.

▸ **Build skills in these areas:**

- Adding layers to a map
- Adding data stored as a spreadsheet or file to a map
- Changing map symbols
- Configuring pop-ups
- Sharing the map as a web application

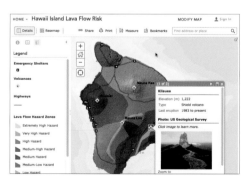

Getting started with ArcGIS Online.

Start Lesson 1

Esri.com/ArcGISBook/Chapter1_Lesson1

Analyze emergency shelter access.

▸ **Overview**

In this lesson, you'll explore the relationship between shelter locations and population in high-hazard lava flow zones. In particular, you'll look for areas where the risk is high and shelter access is poor.

▸ **Build skills in these areas:**

- Using drive-time analysis to evaluate accessibility to locations
- Finding areas that meet specific conditions
- Dissolving many features into a single feature
- Finding how many features in one layer are inside of features from another layer

▸ **What you need:**

- An ArcGIS organizational account
- Estimated time: 2 hours

Start Lesson 2

Esri.com/ArcGISBook/Chapter1_Lesson2

Gulf of Aden

02

Cartography is for Everyone
New ways to make, see, and use maps

Web GIS has changed how people create and engage with geographic information. Online interactive maps form the primary user experience, serving as both the means of creation and the mechanism for delivery. Using maps, you explore locations and access information, discover new relationships, perform editing and analysis, and effectively share your results. In Web GIS, it's all about socializing your map.

SOMALIA

elle

The online mapping revolution

Maps are important. Everyone understands and appreciates good maps. GIS people work with maps every day. Maps provide the basic experience and practical interface for the application of GIS. Maps are also the primary way that GIS users deliver their work.

Maps provide a critical context because they are both analytical and artistic. Maps carry a universal appeal and offer clarity and shape to the world. They enable you to discover and interpret patterns and share your data.

Online maps can be created by virtually anyone using Web GIS—and can be shared with virtually everyone. These maps bring GIS to life and can go with all of us everywhere on our smartphones and tablets.

Make no mistake, traditional printed maps are not going away. They continue to be important because they enable you to quickly get the broad context of a problem or situation. The best printed maps are true works of art that can stir your emotions. There's no comparable large-format document that communicates and organizes such large amounts of information so effectively and so beautifully.

Cartographers using ArcGIS will continue their craft of making astounding print maps that teach and amaze. And this will always be the case. Large-format, printed maps and their digital cousins (like PDFs) will continue to significantly occupy the good work of many mapping professionals. The difference now is that GIS tools have come of age for this level of professional cartography.

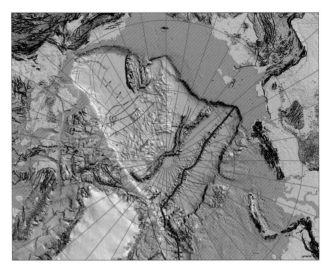

This large-format poster map of tectonic geology, along with sea floor age, and structure appeared in the map gallery at the 2014 Esri International User Conference, and is featured in the cartography category of the 2015 Esri Map Book.

Meanwhile, there is a major online mapping revolution underway, and the implications of this are far-reaching. We all know that consumer maps are ubiquitous on smartphones and the web. Map-based applications regularly rank among the most-used programs on smartphones and mobile devices. Online maps have familiarized millions of people with how to work with maps, and this massive worldwide audience is ready to apply maps in ever more imaginative ways to their work using Web GIS.

GIS maps engage an audience for a purpose

Any map that you make can be saved and shared as a web map—for its intended audience and expected uses. The user experience of your map is determined by the application that you employ. With the ArcGIS platform, users now have more options in designing and implementing purposeful maps, as shown in the apps below.

Emergency responders

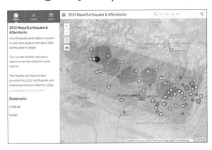

This interactive earthquake map, put up within hours after the devastating 2015 Nepal earthquake, shows the quake epicenters, shaking intensity, and social media updates.

Agricultural managers

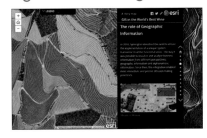

Symington Family Estates is a vintner and port wine house in Portugal that utilizes GIS. They have encapsulated an overview of their GIS work in this story map.

Citizens

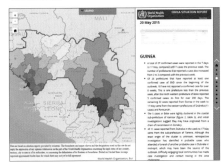

Perhaps no audience is as interested in a map as those trying to survive a dangerous epidemic like Ebola.

Managers and decision makers

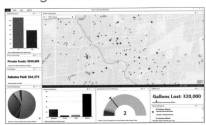

The water conservation dashboard gives a water district executive the ability to monitor water usage in real time.

Citizen scientists

Technology has proven to be vital to Conserve Wildlife Foundation's work over the years protecting rare wildlife species and communicating to its constituents.

Editors

This map interface is an advanced input screen for professional GIS data editors.

What maps can do

Maps can be used to tell stories and apps provide the user experience through which you work with maps and share them. Explore a few examples for how you can leverage your web maps to accomplish your goals.

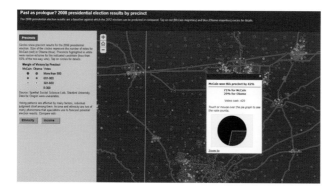

Maps communicate and foster understanding

GIS maps provide windows into useful information. Many kinds of descriptive information can be stored with the map and accessed on demand for individual features in your map. Click on any voting precinct in the map to view its report. For example, you can investigate the relationship between ethnicity and income in presidential voting patterns.

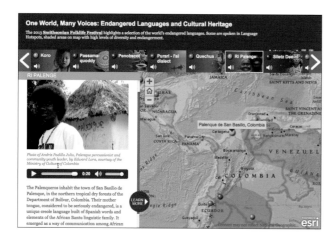

Maps tell stories

Maps provide a powerful way to tell many kinds of stories. ArcGIS story maps make it easy to tell rich, map-based stories in the form of self-contained web apps. These web apps combine intelligent web maps with text, photos, video, and sound to elucidate interesting topics, like this map of endangered languages of the world linked to audio from native speakers.

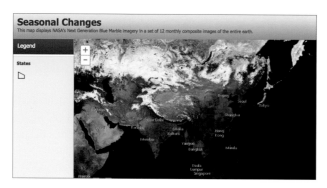

Maps can display dynamic information that changes over time

In GIS, many maps can dynamically display time frames, much like a weather map. This map displays snow cover from NASA's next generation Blue Marble imagery. Click the play button to animate the seasonal imagery over the past 12 months.

Maps help in finding patterns in mountains of data

Although it may seem counterintuitive, sometimes adding thousands or millions of individual features or events helps a more coherent picture to emerge, one that simply couldn't be seen any other way. There can be value in adding mapping detail in order to give clarity to an overall view of your data, and sometimes this is referred to as "data art."

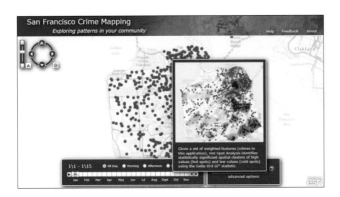

Maps help you perform analysis

Maps can be used to enable geographic analysis. GIS maps combine powerful visualization with a strong analytic and modeling framework. Just as you can use each map layer as a window into information about features, you can use the map as a window into sophisticated analytical tools and results.

Maps can be used to compile data

Sometimes a map is the container or capturing device. This simple map runs on an iPad. A team of arborists in Beverly Hills, California, uses it to add to and update a database of every tree in the city.

The role of web maps

At their heart, web maps are simple

Web maps are online maps created with ArcGIS that provide a way to work and interact with geographic content organized as layers. They are shared on the web and across smartphones and tablets. Each web map contains a reference basemap along with a set of additional data layers, plus tools that work on these layers. The tools can do simple things like open a pop-up window when you click on the map, or more complex things, like perform a spatial analysis and tell you the relative proximity of healthy food options by neighborhood.

At their heart, web maps are simple. Start with a basemap and mash it up with your own data layers. Then add additional tools that support what you want your users to do with your web map: tell stories; perform analytical studies; collect data in the field; or monitor and manage your operations.

Virtually anything you do with GIS can be shared using web maps. And they can go anywhere. Web maps work online and on any smartphone, and along with your supporting GIS work, are accessible anytime.

Web maps are how you deploy your Web GIS. A web map is easy to share with others. You simply provide a hyperlink to the web map you wish to share and embed it on websites or launch it using a wide range of GIS apps.

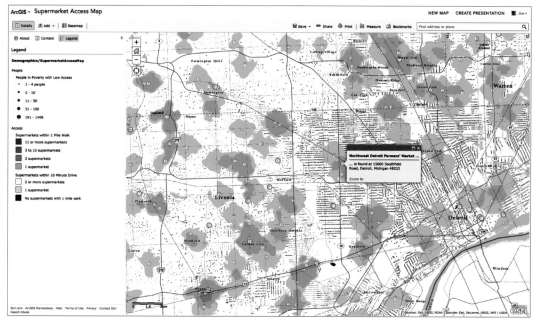

Web maps are how users work with and apply ArcGIS, and can be used anywhere—in web browsers, on smartphones, and in desktop GIS applications. This web map asks a simple question: how many Americans live within a reasonable walking or driving distance to a supermarket or other sources of healthy food?

Make and share a web map

In five easy steps

Anyone can make, share, and use web maps. Let's start by going through a short example. Suppose you want to make a map that allows you to explore the food, architecture, and design destinations for San Diego.

1. Select a basemap and zoom into your city of interest. We'll use the Light Gray Canvas basemap.

2. Add your data layers and specify how each will be symbolized and displayed. In this case, we'll add point layers for specific San Diego destinations, line layers for the trolley lines, and neighborhood polygons.

3. Create pop-ups that enable users to explore the clicked-on features.

4. Save your map, then document it with an eye-catching thumbnail, a good description, and a thoughtful name. Every map, especially a map that you will share, needs a good item description, thumbnail, and name.

5. Share your map by using it to configure a whole range of apps. Once you create your map, you can configure an app that will use your web map for delivery to your audience.

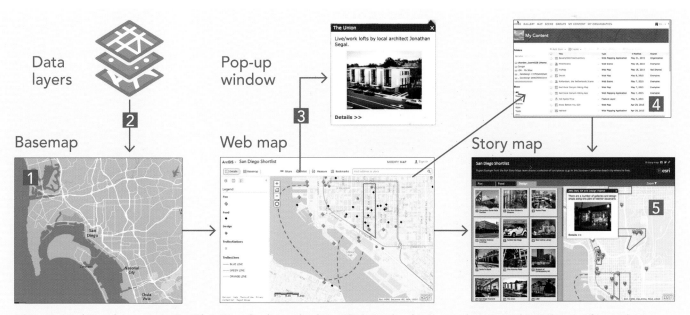

It's easy to share the map you make. Simply share the hyperlink, embed it in a website, or share the configured app.

Basemaps and operational layers

The map mashup is one of the great force multipliers of modern cartography. The ability to easily share and repurpose digital content has allowed each of us to tackle far more ambitious maps than would be possible if we had to work in isolation or start from scratch. The rise of the map mashup expanded cartography, so that anyone could build upon the work of others. Most of the thousands of maps created and shared every day within ArcGIS are mashups—maps that build upon the data, labor, and insights of a larger community. This era of collaborative GIS has empowered everyday citizens to participate in mapping as never before.

It starts with a basemap

In ArcGIS, map authors can readily access beautiful sets of professionally produced basemaps that provide the digital canvas upon which to tell their stories. Each of the Esri basemaps has a theme or focus. Their range serves the need for almost any map type. Whether it's terrain, oceans, roads, or another of the many themes, the right basemap complements your subject and provides the background information critical to establishing its geographic context (locations, features, and labels). Each of the Esri basemaps contains highly accurate and up-to-date information, at multiple zoom levels covering geographic scales from detailed building footprints to the entire planet. Providing data at that level of detail, for all locations on the globe, takes a small army of cartographers and eats up terabytes of data. The good news is that each of us can benefit immediately from those efforts.

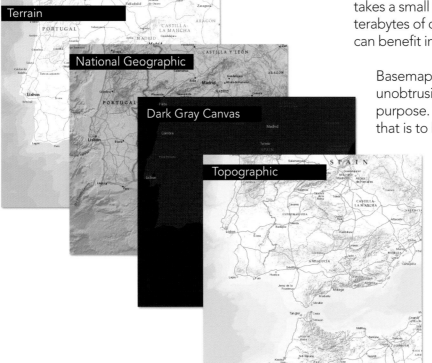

Basemaps seem simple and relatively unobtrusive—and this is precisely their purpose. They should not upstage the content that is to be overlaid on them. "Operational overlays" carry the subject matter of the map and provide the purpose for making any map. An overlay can be anything—air temperature data, life expectancy, the location of oil and gas wells or live traffic conditions. Merging a great basemap with one or more operational overlays forms the heart of the modern web map.

Some map authors are data creators interested in mapping their own data. Many other authors, however, need help finding operational overlays; they know what they want to map but need guidance in finding the data to fully tell the story. Fortunately, ArcGIS provides access to an array of content to use in operational overlays. The ArcGIS community, including Esri, compiles and shares thousands of ready-to-use authoritative datasets, covering everything from historical census data to environmental conditions derived from live sensor networks and stunning earth observations. Finding mappable, interesting geographic data has never been easier.

Blending together ready-to-use basemaps and operational overlays into a live, dynamic map allows you to share geographic content in a simple and concise format.

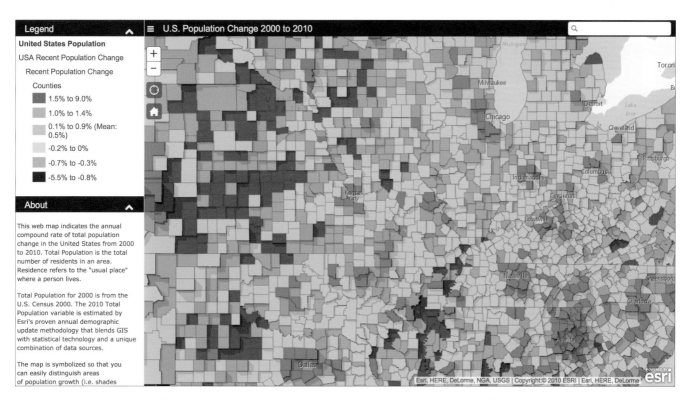

Just imagine trying to understand the subtle spatial patterns of population change for all 3,000-plus US counties by reading a spreadsheet. It would overwhelm anyone. By comparison, a map of that same data can be read and quickly understood by huge numbers of people with almost no training required. This is the power of maps.

Web map properties

Continuous and multiscale

Web maps work across multiple scales. Zoom in to see additional details and gain insight. They're also continuous: they have no edges—you can pan anywhere. Even if you don't have operational data for a particular area, the basemap will still provide reference.

This web map contains the Imagery basemap. As you pan around and then zoom into any spot on the planet, you will find increasing levels of resolution and detail.

Pop-ups

Web maps are windows into a wealth of information. Click on a map location to "pop up" a report and explore the information behind it. Pop-ups help to fit more information into a map since details can emerge on demand. This means a single window into a map can become a window into a world of related information, including charts, images, multimedia files, and other map layers. The ability to link such a wide variety of content to the map has transformed how we think about maps. They've evolved from static containers of data to dynamic information vessels.

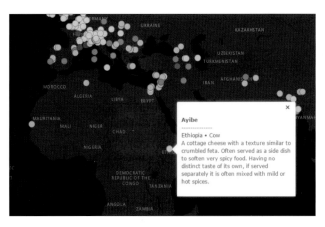

The best pop-ups don't have to be complex. In fact, the simple ones like the one designed into this story map about the World of Cheese are effective because they deliver just what the map author wants you to know about that feature you clicked.

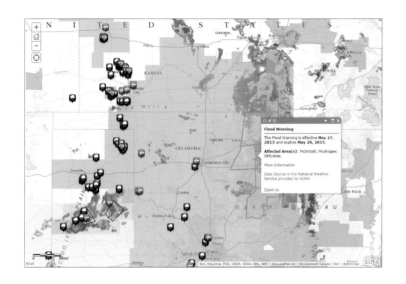

Currency

Your online maps are no longer static. They can be readily and immediately updated because your layers online can contain the latest, most accurate information. When your data changes, the maps that reference that layer are also updated.

This map features live feed layers for severe weather across the United States and Canada (NOAA). Layers can be turned on and off in the Layers panel.

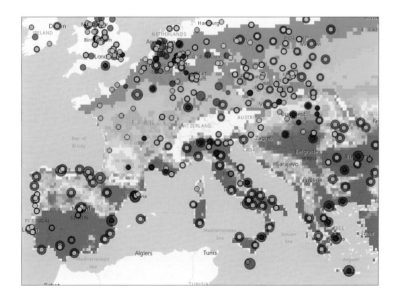

Mashup culture

Your maps can combine more than your own data. You can mash up your rich GIS data with information from other users—in fact, whatever is useful and relevant to your objectives from the entire world of GIS users.

This web map on heat wave risk mashes up data from the European Environment Agency (EEA), EUROSTAT, and the Swiss Institute for Atmospheric and Climate Science. The original map was presented in the EEA report "Urban Adaptation to Climate Change in Europe."

New smart mapping workflows

Your world is full of data, and maps help you to make sense of it. There is a growing need to turn geographic data into compelling maps. All users want to create beautiful, interactive maps and infographics with live data, easily and with confidence. The smart mapping mission is to provide a new kind of strong "cartographic artificial intelligence" that enables virtually anyone to visually analyze, create, and share professional quality maps in just a few minutes, with minimal mapping knowledge or software skills.

Smart mapping is designed to give ArcGIS users the confidence and ability to quickly make maps that are visually pleasing and effective. Cartographic expertise is "baked" into ArcGIS, meaning it's part of the fundamental user experience of using ArcGIS. The map results that you see in front of you are driven by the nature of the data itself, the kind of map you want to create, and the kind of story you want to tell.

By taking the guesswork out of all of the settings and choices that you could conceivably tweak, your initial map results are cartographically appropriate and look wonderful. You can always change things at will, which you'll undoubtedly do as you gain more experience, but smart mapping gets you to something effective very quickly. You spend less time iterating and wrangling your maps into fulfilling your intention.

The point is not to take control away from map authors or dumb down the map-authoring experience, but to be smarter about how all of the initial parameters of the map (color, scale,

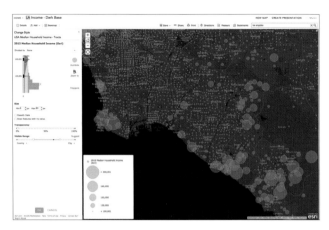

The best thematic maps apply thoughtful analysis of the underlying data to a set of map parameters designed to bring focus and clarity to the topic. Here, the author has positioned the handles controlling symbol size (running beside the histogram) to emphasize areas with household incomes over $100,000. Great maps relate the data back to the real world, using visual cues that immediately highlight the message you want to convey.

styling, etc.) are established. For example, each of the Esri basemaps (Streets, Dark Gray Canvas, Topographic, etc.) were paired with several multi-hue color schemes that can be used as the defaults for your operational layers. This way you know your map will look good right out of the box without needing any adjustments.

Mapping professionals still have full control and the ability to extend the default capabilities to create unique customizations and truly exquisite, publication-quality cartography.

Thought Leader: Jim Herries

Map design is about drawing your audience into the story you're telling

The most valuable maps are information products. They are visually interesting the very first time you see them, and they reward you with additional information as you look around the map and zoom to an area you know. When you touch the map, it responds by giving you details about the thing you touched—touch a store and it tells you this year's sales to date with a chart of the previous three years' sales. Maps are interactive, rewarding experiences, and not just pretty pictures.

Great maps don't just happen automatically, though. You have to put a little bit of yourself into the effort, just like a great resume, which starts out as a template but requires your information— your data—as well as your interpretation to make it really sing. The data you are mapping won't tell its story without your help. Once you see the patterns emerging in the map, you can start emphasizing what's important, and de-emphasizing everything else.

Try to always make "beautiful" maps. By that I mean effective ones that are clear on first opening but that also entice users of all levels to drill in, explore, interrogate, and learn.

Start with the final result you have in mind and work backwards. To paraphrase Roger Tomlinson, one of the fathers of GIS in the 1960s, you've got to know what you want to get out of a GIS in order to know what to put into it. Clicking aimlessly leads to a world of hurt. Have a clear idea of what you want to produce, explain, or monitor. Next, get some test data, and then "have a play."

Jim Herries is an applied geographer with Esri in Redlands, California. Every day, he makes maps and helps others make their maps better by eliminating the noise and increasing the signal.

 Watch a video: Map Makeovers: How to Make Your Map Great
Esri.com/ArcGISBook/Chapter2_Video

ArcGIS for cartographers
The art of mapmaking

ArcGIS for Desktop, including the new ArcGIS Pro application, provides capabilities that enable everyone to make truly excellent maps, including support for highly sophisticated mapping workflows employed by professional cartographers. Desktop includes tools for rich data compilation, for importing data from a multitude of publication formats, and for integrating this data with your own data to create consistent, accurate, and beautiful cartographic products for both printed maps and online maps.

Two key applications available in Desktop provide advanced mapping capabilities. ArcMap has been the workhorse application for serious cartographic production to create print maps and online maps with advanced labeling and impressive cartographic symbols and representations. It is used daily by hundreds of thousands of GIS users worldwide. The new arrival is the modern ArcGIS Pro application, which builds on the tradition for great mapping and adds things like advanced 3D scenes.

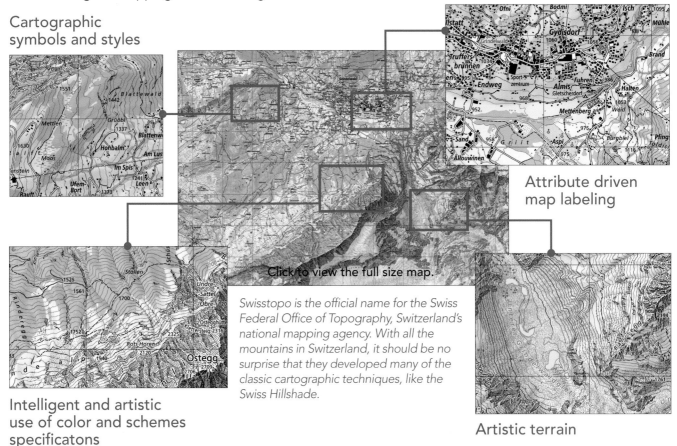

Cartographic symbols and styles

Attribute driven map labeling

Click to view the full size map.

Intelligent and artistic use of color and schemes specificatons

Swisstopo is the official name for the Swiss Federal Office of Topography, Switzerland's national mapping agency. With all the mountains in Switzerland, it should be no surprise that they developed many of the classic cartographic techniques, like the Swiss Hillshade.

Artistic terrain

Representing elevation and terrain
Using ArcGIS for Desktop

Professional mapmakers, who often work at creating map series products for both print publications and their own online basemaps, create and manage their map designs using the desktop applications ArcMap and ArcGIS Pro. They design each map as an ordered series of map layers that get overlayed and combined with other layers, and then symbolized on the final map.

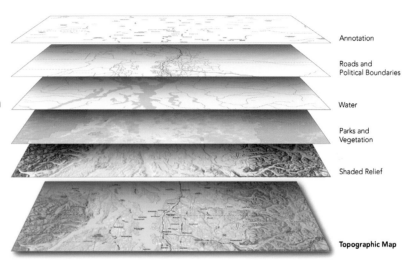

Annotation

Roads and Political Boundaries

Water

Parks and Vegetation

Shaded Relief

Topographic Map

This is an example of how the desktop tools come into play, carrying out the cartographic heavy lifting. Almost every basemap contains a terrain layer, often represented as relief and contours. The terrain layer is literally the foundation for these basemaps. A key requirement for cartographers who create them is to apply useful methods for making detailed and artistic hillshades of their own data for use in their basemaps. Cartographers can download a set of tools including useful terrain mapping techniques, for representing terrain under different lighting conditions.

Terrain tools
One of the highlights of Terrain Tools is the new Cluster Hillshade that enables you to make spectacularly detailed and artistic hillshades with your own data. This is just about as close as you can get with an automated process to classic hand-drawn hillshading—and all from just a Digital Elevation Model input and the click of a mouse. Download Terrain Tools here for working with ArcGIS Desktop.

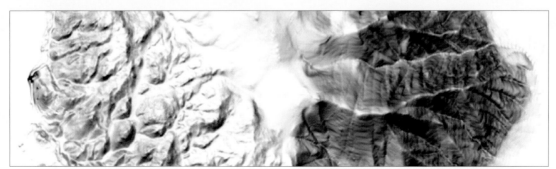

Examples of widely used methods for terrain representation are available as a toolkit for Desktop users.

Quickstart

Get inspired and learn current mapping techniques using curated selections of exemplary cartography at the Maps We Love website

What makes a good map? How can you engage people with a map? How do you make a map that offers unexpected insights and captivating appeal? We have been working on something at Esri that we hope will answer these questions: Maps We Love.

Maps We Love is an ongoing project where you will see the best of what's possible with ArcGIS. This is where you come for the inspiration, ideas, and information you need to turn your data into brilliant maps. We give you a behind-the-scenes look at important steps, plus resources (lots of links) so you can dig deeper into these topics. Maps We Love is designed to demystify mapping, to give you the confidence and assurance that you can make maps.

Go to Maps We Love

www.MapsWeLove.com

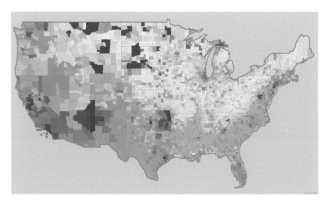

US Minority Populations

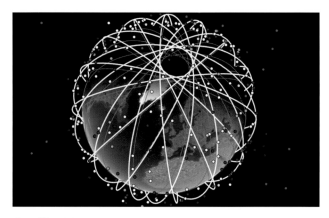

Satellite Map

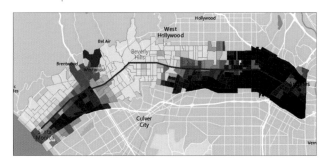

 Watch a smart mapping video demo
Esri.com/ArcGISBook/Chapter2_Video1

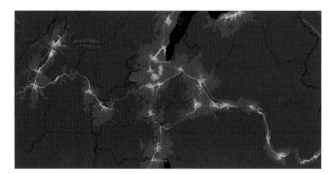

Highway Access in Europe

Learn ArcGIS Lesson

Do a complete desktop analysis and mapping project

▸ **Overview**

The Amazon rainforest spans nine countries and millions of square kilometers, making it the largest tropical rainforest in the world. Since the 1960s, the rainforest has undergone significant deforestation. Current estimates indicate only about 80 percent of the original rainforest remains.

One of the most deforested regions is the Brazilian state of Rondônia. In 2011, a Brazilian judge prohibited the construction of a road that would have traveled through a large stretch of protected land in Rondônia, potentially causing even more deforestation.

In this project, you'll use ArcMap to predict how much deforestation was prevented by prohibiting a proposed road. First, you'll find the study area. Then, you'll compare roads and deforestation to determine the pattern of existing deforestation, before applying your findings to the proposed road. Lastly, you'll communicate the results to others.

▸ **Build skills in these areas:**
- Adding data to a map
- Analyzing relationships between data
- Digitizing features
- Symbolizing and organizing data

▸ **What you need:**
- ArcMap (see page 15)
- Estimated time: two hours

Start Lesson

Esri.com/ArcGISBook/Chapter2_Lesson

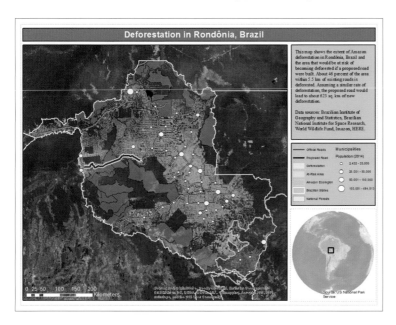

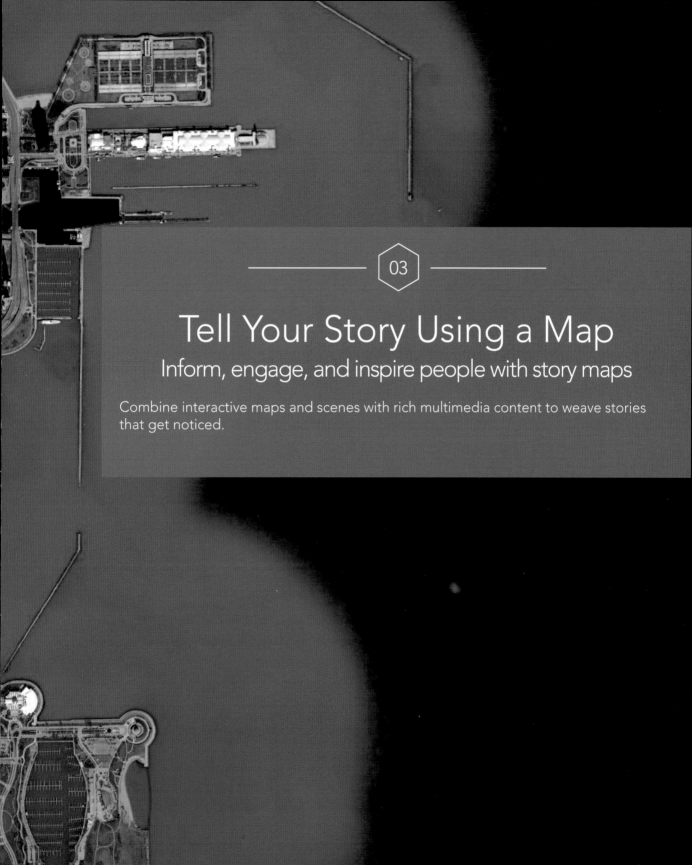

03

Tell Your Story Using a Map

Inform, engage, and inspire people with story maps

Combine interactive maps and scenes with rich multimedia content to weave stories that get noticed.

Story maps
The fusion of maps and stories comes to life

Storytellers often turn to maps to expand and illuminate their words. Maps are the visual representation of where events happen. As such, maps and stories complement each other, but until recently they have existed more as side-by-side products and not one thing. The big idea of this chapter is that with story maps they can now be one thing.

Story maps use geography as a means of organizing and presenting information. They tell the story of a place, event, issue, trend, or pattern in a geographic context. They combine interactive maps with other rich content—text, photos, video, and audio—within user experiences that are basic and intuitive. While many story maps are designed for general, non-technical audiences, some story maps can also serve highly specialized audiences. They use the tools of GIS, and often present the results of spatial analysis, but don't require their users to have any special knowledge or skills in GIS. This has resulted in a veritable explosion of story maps. (Go to www.TheArcGISBook.com to see them come alive.) As you click through to the various story maps linked in the chapter, allow yourself the freedom to immerse yourself in the narrative. These are information products that reward exploration.

With today's cloud-based mapping platform, the fusion of maps and stories has come of age. Consider the work of a venerable publication that has moved deftly into web delivery of their stories with Smithsonian.com.

"The Age of Humans" is an important article published by Smithsonian Magazine. Readers of the online edition are treated to this embedded story map that brings together satellite imagery and other datasets to create an atlas of humanity's influence on the planet. Zoom in to see regional effects, or click through some selected examples, including a gallery of international extinctions and a virtual tour of the world's protected sites.

The world of story maps

A gallery of exceptional examples from around the globe

The international Esri user community fuels a prolific information-creation engine, driving into view on the ArcGIS platform the most authoritative work on the world's most pressing and serious issues. The imaginative use of story maps and the live examples featured on this page and in the Esri-curated Story Maps Gallery are designed to show the range of ways that such narratives convey information.

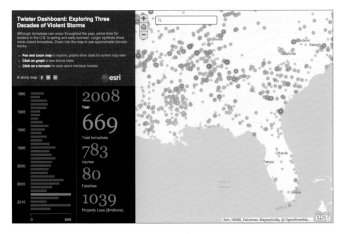

The Twister Dashboard beautifully synthesizes 30 years of tornado data.

Geography Bee features exquisite insect photography coupled with global species location overviews.

The world is a crowded place, with more than 7 billion people on the planet as of 2014. About half of this population lives in a megacity—a metropolitan area with 10 million people or more.

This map tells the story of rapid highway development and love of the car in China.

Who creates story maps?

For the people, by the people

Storytelling carries the potential to affect change, influence opinion, create awareness, raise the alarm, and get out the news. The answer to the question of who authors story maps is anybody—any individual or group that wants to communicate effectively, including you. Here are a few examples, created by users just like you, to spur your imagination.

News organizations

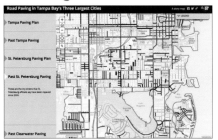

ABC TV in Tampa, Florida, uses story maps to drive traffic to its online news articles when the story has a particularly spatial angle, like this one about street paving.

Nonprofits and NGOs

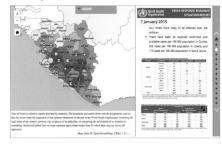

In this story map, the World Health Organization maintains daily Situation Reports on the 2014–2015 Ebola outbreak.

Scientists

Researchers at VHL University of Applied Sciences tell the story of alarming declines in frog and newt populations in the Netherlands.

Tourism boards

Story maps are a way to tell people about a tourist destination like Palm Springs. This Shortlist is the definitive guide to the desert resort, in part because it is kept up to date.

Citizen bloggers

To express their fascination for a certain place or region, like Southern California's Inland Empire, some post their own weird and wonderful interpretation of what intrigues them.

Public art

The people of Ireland take their roadside art seriously. This impressive story map serves as both a guide and a justification for the public funding of such works.

Thought Leader: Allen Carroll

Why maps are so interesting

For most people sight is the dominant sense, so when it comes to information delivery, most like it served visually. One way to think about it is to consider that as information publishers we actually have relatively few ways to organize information. We can alphabetize it, but that's not very much fun. We can arrange it by time, chronologically, but that has its limitations. We can organize knowledge taxonomically by category or hierarchically in some kind of ranking. And then we come to spatial organization, the system that arranges things by where they are. This one offers unique insights and the potential to visualize information. Organizing by location is a particularly interesting and useful way to marshal information.

Another reason why so many relate to maps and geography is that we have no choice but to think and see spatially. We have to make sense of our surroundings and navigate through our world. Maps make sense of things. They lend order to complex environments and they reveal patterns and relationships.

Maps can also be quite beautiful. They stimulate both sides of our brain: the right side that's intuitive and aesthetic, and the left side that's rational and analytical. Maps are this wonderful combination of both. It's this neat marriage of utility and beauty that I find alluring.

For more than two decades, Allen Carroll told stories with maps at National Geographic. As the Society's chief cartographer, he participated in the creation of dozens of wall maps, atlases, globes, and cartographic websites. Today he leads the Esri Story Maps team, which uses state-of-the-art GIS technology, combined with digital media, to bring maps to life in new ways.

Watch a video on storytelling and information design
Esri.com/ArcGISBook/Chapter3_Video

Maps tell stories
What kinds of stories can you tell?

Describing places

Some maps do the very basic work of describing places. These are the maps we use to navigate the world. Anyone planning to summit Mt. Langley in California's eastern Sierras will benefit by consulting this story map.

Comparing data

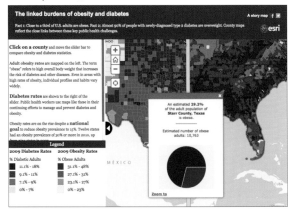

Fact 1: Close to a third of US adults are obese. Fact 2: Almost 90% of people with newly diagnosed type 2 diabetes are overweight. This county-level story map reflects the close links between these key public health challenges.

Revealing patterns

This story map explores correlations between officer-involved shootings and race, poverty, and crime.

Presenting narratives

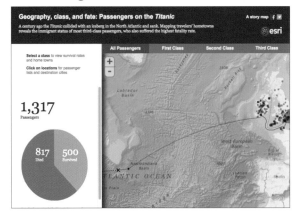

Passenger rolls reveal that third-class passengers suffered the highest fatality rate by far when the Titanic sank.

Recounting history

With a spatial analysis that examines viewpoints of both Union and Confederate commanders at key moments in the Battle of Gettysburg, this story map reveals the critical role that geography played in the decisive US Civil War battle.

Celebrating the world

The world is a strange and wonderful canvas of cultural differences. This story map, depicting unusual sports around the globe, leverages live embedded video for a comical effect.

Breaking news

With each phone a sensor and every Twitter post a data point, it's a crowdsourcing bonanza for this astronomical effort by the Centre of Geographic Sciences (COGS).

Depicting change

A spyglass compares a historical map from the David Rumsey Collection with a current satellite image; published by Smithsonian.com to show the city's growth.

Quickstart

Combine your maps and customize interactive apps to tell a story

Things to consider when creating a story map:

▸ **Think about your purpose and audience**
Your first step is to think about what you want to communicate with your story map and what your purpose or goal is in telling the story. Who is your audience? Are you aiming your story at the public at large, or a more focused audience, like stakeholders, supporters, or specialists who would be willing to explore and learn about something in more depth?

▸ **Spark your imagination**
Go to the Story Maps Gallery to see some examples handpicked by the Esri Story Maps team to inspire you and highlight creative approaches. You can filter and search the gallery to check out how authors have handled subjects and information that may well be similar to yours. Explore. Get a gut feel for what makes a good story.

▸ **Choose a story map application template**
Go to the Story Maps Apps to browse the application templates and choose the best one for your story map project. Each app lets you deliver a specific user experience to your audience. There are apps for map-based tours, collections of points of interest, in-depth narratives, presenting multiple maps, etc.

▸ **Follow the instructions for the application template you chose**
See the Tutorial tab for the story map application template you chose for instructions on how to proceed. For example, here is the Tutorial for the Story Map Journal application template.

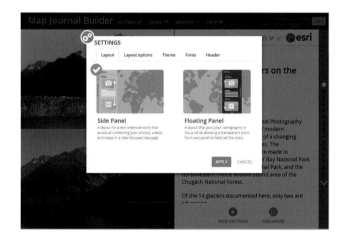

▸ **Make your story map go live and promote it**
When you've finished, you simply share your story map to launch it and make it go live. You can share it publicly or restrict it so it can be accessed only by people in your organization. To promote the story map to your audience, you can add links to it, embed it into your website, write a blog post about it, and share it on social media.

Learn ArcGIS Lessons

Make your own story map with a smartphone

In these two lessons, you'll go through the process of using an online photo gallery, or a gallery created with your own smartphone, to create a personal story map. We encourage you to use your own GPS-enabled device, but if you don't have one or just want to do this as quickly as possible, then we have set up a Flickr account where you can grab the photos that we used and go from there.

▶ Overview

Today's smartphones and other GPS-enabled cameras capture (with surprisingly high accuracy) the location where each photo was taken. The geographic coordinates, called "geotags," stay associated with the photos when you upload them to Flickr or another hosting environment, and can be used to quickly assemble a Photo Map Tour like this one in ArcGIS Online.

This project allows you to use your creativity to its fullest. In the first example, use a series of photographs taken along The Strand, a three-mile strip of beach in Los Angeles County renowned for its culture and beauty; or you can choose to recreate the geoportfolio of a photography student. In either case, you can use the provided photos or take your own.

Option 2: Make a traditional map tour.

▶ Build skills in these areas:
- Using the built-in GPS features of a smartphone or camera to geotag pictures
- Using the Story Map Tour builder
- Image handling, photography (optional)

▶ What you need:
- Computer with a web browser
- An ArcGIS Online organizational account or an ArcGIS Online public account (see page 15)
- Time estimate: 30 minutes to an hour

| Start Lesson 1 | Start Lesson 2 |

Esri.com/ArcGISBook/ Chapter3_Lesson1   Esri.com/ArcGISBook/ Chapter3_Lesson2

Option 1: Make a geoportfolio.

04

Great Maps Need Great Data
Creating and using authoritative geographic data

ArcGIS Online is rapidly emerging as the platform of choice for the creation and dissemination of authoritative geographic data content. This Living Atlas of the world is a highly active network of contributors and curators whose output is accessed billions of times monthly. This chapter explains how this unique data ecosystem works, how to access that data, and how to contribute your own piece to the puzzle.

The Living Atlas
The ArcGIS platform provides rich content

The Living Atlas of the World is a treasure trove of information, a dynamic collection of thousands of maps, data, imagery, tools, and apps produced by ArcGIS users worldwide (and by Esri and its partners). Think of it as the curated subset of ArcGIS Online as a whole, organized by the ArcGIS community. This deep and definitive catalog of information awaits you. And that's the big idea of this chapter, that you can combine content from this repository with your own data to create powerful new maps and applications.

The Living Atlas represents the collective work of the entire mapping community—the people who use the ArcGIS platform as the system of record for their work. As such, it is fast emerging as the extensive and most authoritative source of geographically referenced information on the planet.

Hunting down good data used to involve a lot of work just to get a GIS project started. These days, using ready-made basemaps and authenticated data from ArcGIS Online, GIS analysts are able to spend more time thinking analytically, which really gets to the heart of what makes Web GIS work.

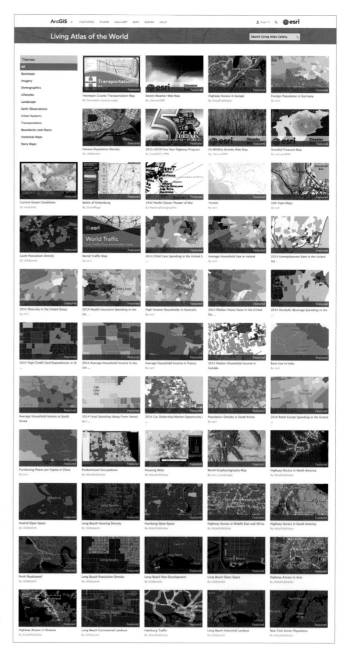

ArcGIS includes a Living Atlas of the World with beautiful and authoritative maps on thousands of topics. Explore maps and data from Esri and other organizations and combine them with your own data to create new maps and applications.

The ArcGIS data community

A global network for creating and sharing authoritative geographic information resources

The mission of every GIS organization is to perform specific functions within its jurisdiction. Each of these departments, groups, or agencies is committed to building key authoritative data layers to support its work. This work includes the compilation of foundational data layers as well as standard basemap layers for their geographies and applications.

For such organizations—and they are myriad in local, regional, state, and national levels around the globe—this information has served as the basis for all of their comprehensive GIS applications. During the early days of GIS, the compilation of these data layers was one of the primary tasks of each organization. As this data

was developed, GIS data developers were able to leverage their information resources in various kinds of GIS applications that extend their own work and help their constituents.

The result is that all these different agencies have created data that is considered, in legal terms, a system of record maintained to support their mandated domain. The pace of migrating this data into Web GIS is increasing, and we are now seeing many contributions coming online that fill in gaps for the entire world. The result is a continuous coverage of geographic information worldwide—the Web GIS of the world.

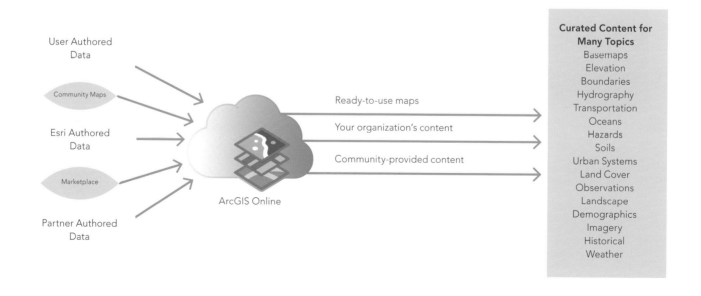

What kind of data is available?

Definitive, authoritative

Imagery

Image layers enable you to view recent, high-resolution imagery for most of the world; multi-spectral imagery of the planet updated daily; and near real-time imagery for parts of the world affected by major events, such as natural disasters.

The waters off the coast of Greenland are covered in a beautiful, swirly pattern of sea ice in this true-color image taken by NASA's Aqua satellite on Aug. 18, 2014.

Basemaps

Basemaps provide reference maps for the world and the context for your work. Built from the best available data from the ArcGIS community of reliable data providers, these maps are presented in multiple cartographic styles.

Designed for marine GIS applications, the Ocean basemap includes bathymetry, derived depths, and surface and subsurface feature names.

Demographics and Lifestyles

Demographics and Lifestyles maps—of the United States and more than 120 other countries—include recent information about total population, family size, household income, spending, and much more.

This map of Singapore emphasizes areas with the highest population density (more than 50,000 persons per square kilometer).

Boundaries and Places

Many places are logically defined by a boundary. Boundaries and Places map layers describe these areas at many levels of geography, including countries, administrative areas, postal codes, and more.

This layer presents time zones of the world. The layer symbolizes the time zones in relation to the Prime Meridian—the line of 0 degrees longitude passing through Greenwich, England.

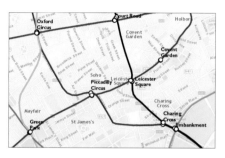

Transportation

This category contains maps and layers that describe the systems that people use to move between places. It includes a variety of global, national, and local maps on various topics from infrastructure projects to rest areas. Some of these layers are dynamic, such as the live World Traffic map, which is updated every few minutes with data on traffic incidents and congestion.

This map shows the Tube routes and stations in London.

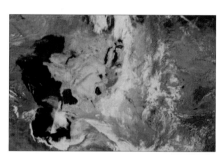

Earth Observations

Earth Observations maps and layers are collected from sensors on the ground and satellites in space. They describe our planet's current conditions, from earthquakes and fires to severe weather and hurricanes, and they enable us to see changes over time.

Taken from a NOAA weather data service, this image is from a real-time feed of precipitation over Florida.

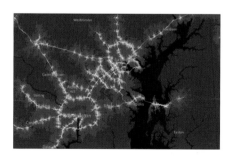

Urban Systems

These layers depict data about human activity in the built world and its economic activities and include things like utility infrastructures, parcel boundaries, 3D cityscapes, housing, and employment statistics.

Areas that are within 10 minutes of a freeway exit are emphasized on this map to give an indication of how accessible neighborhoods are by highway.

Historical Maps

This collection includes scanned raster maps presented as both static map layers and dynamic image layers. These layers can be viewed individually as a basemap or displayed with a current basemap for comparison purposes.

A classic world map published in December 1922 shows the world's political boundaries established following WWI.

Basemaps
The setting for your story

A basemap provides a reference map for your world and a context for the content you want to display in a map. When you create a new map, you can choose which basemap you want to use. Change the basemap of your current map at any time by choosing from the basemap gallery or using your own basemap.

The evolution of basemaps has quietly changed life for the everyday mapping professional. They make it easy to create most maps. Billions of ArcGIS maps utilizing these basemaps are created and shared every week. There are several key concepts to understand about basemaps. They are multiscale and continuous and provide global coverage.

Multiscale
This means that as you zoom into or out of a map, the features and detail that you see change. The ArcGIS basemap collection is continuous in scale. Zoom from the entire planet into the details of your neighborhood and down to a single property parcel.

Global coverage
These maps cover the entire surface of the earth. Basemap coverage and levels of detail are improving each day as more data is added to the system.

Continuous
This means they never stop; basemaps wrap around the surface of the earth.

World Imagery provides satellite and aerial, cloud-free imagery in natural color, at one meter or less, of many parts of the world and lower resolution imagery worldwide.

With exactly the same imagery as the World Imagery basemap, this map includes political boundaries and place names for reference purposes.

This comprehensive street map includes highways, major roads, minor roads, railways, water features, cities, parks, landmarks, building footprints, and administrative boundaries overlaid on shaded relief.

This basemap shows cities, water features, physiographic features, parks, landmarks, highways, roads, railways, airports, and administrative boundaries overlaid on land cover and shaded relief for added context.

Dark Gray Canvas

This dark basemap supports bright colors, creating a visually compelling map graphic that helps your reader see the patterns intended by allowing your data to come to the foreground.

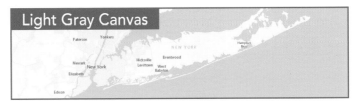

Light Gray Canvas

Like its dark counterpart, this basemap supports strong colors and labels against a neutral, informative backdrop. The canvas basemaps leave room for your operational layers to shine.

National Geographic

The map was developed by National Geographic and Esri and reflects the distinctive National Geographic cartographic style in a multiscale reference map of the world.

Ocean

The Ocean basemap (showing coastal regions and the ocean seafloor) is used by marine GIS professionals and as a reference map by others in the oceans and maritime community.

Terrain with Labels

This basemap features elevations as shaded relief, bathymetry, and coastal water features that provide a neutral background with political boundaries, and place names for reference purposes. This is often relevant for natural resources applications.

OpenStreetMap

OpenStreetMap (OSM) is the open collaborative project to create a free editable map of the world. Volunteers gather location data using GPS, local knowledge, and other free sources of information.

USA Topo Maps

This set of maps provides a very useful basemap for a variety of applications, particularly in rural areas where topographic maps provide unique detail and features from other basemaps.

USGS National Map

This composite topographic basemap of the US, is provided by USGS. It includes contours, shaded relief, woodland and urban tint, along with vector layers, such as governmental unit boundaries, hydrography, structures, and transportation.

Demographics

This data about populations includes the basics, like age and ethnicity, but also peoples' wealth and health, their spending habits, and their politics. ArcGIS includes many hundreds of demographic variables that are accessible as maps, reports, and raw data that you can use to enrich your own maps.

The idea of data enrichment means that you can associate or append demographics to your local geography. This ability to combine your existing data with demographic variables specific to the problem being studied has opened a whole new avenue for everyone, not just consumer marketers, but epidemiologists, political scientists and sociologists and really any professional who wants to better understand a certain segment of human population.

Demographers often want to understand populations not only in the current time but also into the future. How will a given population group change over time? The art of forecasting current-year estimates based on the decennial US Census, for example, is something that is very carefully conducted by the demographic experts at Esri.

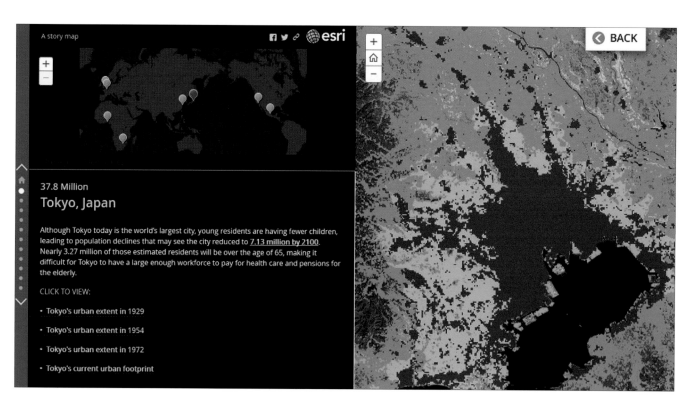

Updated Demographics

Accurate current-year estimates and five-year projections for US demographics, including households, income, and housing.

Census and ACS

Census and American Community Survey (ACS) data used to analyze the impact of population changes on services and sites.

Tapestry Segmentation

Detailed descriptions of residential neighborhoods, including demographics, lifestyle data, and economic factors divided into 67 segments.

Consumer Spending

Data about products and services consumers are buying. Includes apparel, food and beverage, entertainment, and household goods and services.

Market Potential

Includes thousands of items that consumers want. The Market Potential Index (MPI) measures consumer behaviors by area compared to the US average.

Retail Marketplace

Direct comparison between retail sales and consumer spending by industry. Measures the gap between supply and demand.

Business Data

Business Locations and Business Summary data from Dun & Bradstreet. Provides sales, employee information, industry classification, and more.

Major Shopping Centers

Statistics for thousands of major shopping centers, collected by the Directory of Major Malls. Includes name, total sales, and more.

Crime Indexes

Statistics about major categories of personal and property crime. Includes information about assault, burglary, and more.

Traffic Counts

Peak and low traffic volume of vehicles that cross a certain point or street location. Contains more than one million points.

Global Datasets

Esri Demographics includes global datasets that range from population and lifestyles to consumer spending and traffic counts.

Global Demographics

Recent demographics about total population, family size, household income, education, marital status, household type, unemployment, and more.

Global Spending

Total amount spent and amount spent per capita for categories such as food, clothing, household, medical, electronics, and more.

Opening data to the world of possibilities

Open Data allows organizations to use the ArcGIS platform to provide the public with open access to their geospatial data. Organizations use ArcGIS Online to create their own website and specify Open Data groups to share specific items. The general public can use Open Data sites to search by topic or location, download data in multiple formats, and view data on an interactive map and in a table. Here are some examples.

ArcGIS Open Data community

ArcGIS Open Data community provides direct access to thousands of open government datasets. Citizens can search, download, filter, and visualize this data through their web browser or mobile device.

Data Driven Detroit (D3)

D3 believes that direct and practical use of data by grassroots leaders and public officials promotes thoughtful community building and effective policymaking. As a "one-stop-shop" for data about the city and metro area, D3 provides unprecedented opportunities for collaboration and capacity building in Southeast Michigan.

City of Halifax, Nova Scotia

As part of Halifax's commitment to improving citizen engagement and enhancing transparency and accountability to its residents, the municipality provides public access to its datasets. The Halifax Open Data Catalogue is now a permanent service provided for citizens and businesses alike.

Open Charlotte

Charlotte, North Carolina's open data site offers all the city's data, providing citizens with a higher degree of visibility into public services, which can foster trust in local government.

Imagery

At the most basic level, imagery is simply pictures of the earth. Imagery can be immediate or taken across multiple time spans enabling us to measure and monitor change. Every image contains massive amounts of information and can be one of the most immediate ways to collect data.

When it's integrated with GIS, imagery encompasses a broad collection of data about our world in the form of pictures from above—taken by satellites from space, aircraft flying over our cities, and collected by other sensors. Imagery represents the earth in digital pictures composed of millions of pixels. Satellite and aerial images are geo-referenced pictures that overlay focused areas of our planet.

Because imagery sees the earth in unique ways, this enables us to both view and analyze our world using multiple perspectives. Depending on the satellite's sensors, imagery can provide access to both visible light as well as to invisible aspects of the electromagnetic spectrum. This enables us to interpret what we can't see with the naked eye. We can visibly observe the presence or absence of water, classes of land cover and urbanization, the occurrence of certain minerals, human disturbance, vegetation health, changes in ice and water coverage, and a multitude of other factors. Imagery even enables us to automate the generation of 3D views of our planet.

Because the imagery collection is immediate, it enables us to monitor and measure change over time.

Photographic

Aerial photography, historically on film, has gone digital. Still and video imagery from drones is on the rise. After this May, 2014 Oklahoma tornado event, updated imagery for the scene appeared online within 24 hours.

Satellite

The extent of satellite imagery (known as Landsat) has never been better. It can be used for mapping crops and vegetation, gold and oil exploration, and urbanization, to name just a few examples.

Multispectral

Electronic sensors in satellites and planes detect more than the human eye—information in the form of spectral bands. Once a band is captured as an image by a sensor, it can be displayed using the colors we see.

Landscapes

Landscape analysis layers

Landscape analysis underpins our efforts to plan land use, engage in natural resource management, and better understand our relationship with the environment. Esri has taken the best available data from many public data sources and provided the content in an easy-to-use GIS collection of datasets.

The map layers in this group provide information about natural systems, plants and animals, and the impacts and implications of human use of those resources that define the landscape of the United States and the rest of the world.

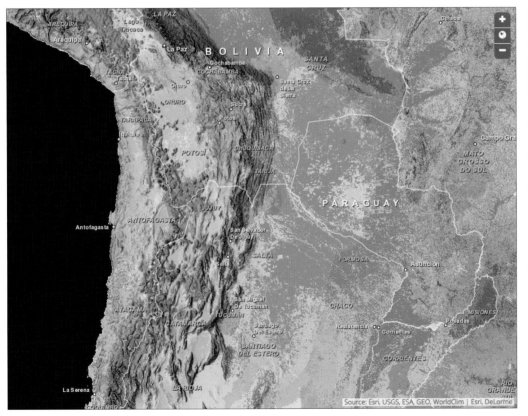

Ecological Land Units (ELUs) portray ecological and physiographic information about the earth. They provide an accounting framework for assessments of carbon storage and soil formation, and of important risk factors such as environmental degradation. ELUs also lend themselves to the study of ecological diversity, rarity, and evolutionary isolation. For example, we can identify the most diverse landscapes in terms of unique ecological land features.

Thought Leader: Richard Saul Wurman
A map is a pattern made understandable

There's a notion that the more you put on the map, the better the map, but there's a case where the opposite is true. Put two patterns together and you'll discover a third. Pile on too much and you can't discover a pattern at all.

The simple phrase "understanding precedes action" was an off-the-cuff remark I made that resonated as a truism, so the phrase stuck. Here's an example that illustrates why it's an important idea.

As cities incur more traffic, they add more freeways and highways. Yet does that actually solve the problem? Or does it spur the purchase of more cars, further crowding on freeways, while we consume fuel and generate more pollution? Adding more lanes only invites more traffic. The problem wasn't understood, but action was taken anyway.

Understanding precedes action. This is at the heart of the Urban Observatory, a longtime dream recently realized with the help of my friends at Esri. It's a simple idea. Yet simple is not necessarily reductive or dumbed down. In fact, it can be edifying. That's how I see it. And GIS is the key to this kingdom. It transforms mapping into a universal language and gives you the opportunity to ask questions and find answers visually. In fact, GIS allows us to ask better questions.

Richard Saul Wurman is an American architect and graphic designer. He has written and designed over 80 books and co-founded the TED (Technology, Entertainment, and Design) conference, as well as numerous other conferences. With a lifelong passion for creating understanding, Wurman has an extensive interest in maps, cartography, and design, culminating in his collaboration with Esri to create the Urban Observatory, which takes advantage of GIS as an integrative platform.

 Watch a talk by
Richard Saul Wurman
Esri.com/ArcGISBook/Chapter4_Video1

Explore the Urban Observatory

QuickStart

Using and contributing to the GIS data ecosystem

Browse the Living Atlas

ArcGIS includes a Living Atlas of the World with
beautiful and authoritative maps and layers on
hundreds of topics. These maps are shared by Esri,
our partners, and members of the ArcGIS user
community. You can help enrich the Living Atlas
by sharing your maps and apps. The Living Atlas is
curated so you, and others, can count on finding
high quality information for your ArcGIS applications.

Contribute to the Living Atlas

To share your content items with the ArcGIS
community, please nominate your publicly shared
ArcGIS Online maps and apps for review by our
curators.

Read a blog about how to nominate your content
for consideration.
Esri.com/ArcGISBook/LivingAtlasBlog.

*The collection of maps, intelligent map layers,
imagery, tools, and apps built by ArcGIS users
worldwide and by Esri and its partners is accessible at
Esri.com/ArcGISBook/LivingAtlasGallery.*

Unlock Earth's Secrets with Landsat Imagery

 **Watch a video about contributing
your data to the Living Atlas**

Esri.com/ArcGISBook/LivingAtlasVideo.

*Landsat sees Earth in a unique way. It takes images of
every location in the world to reveal hidden patterns
in everything from volcanic activity to urban sprawl.*

Learn ArcGIS Lesson

Leverage free data to make an app that is engaging, educational, and informative

Red Rock Canyon National Conservation Area, located just west of Las Vegas, welcomes more than one million visitors a year. It offers a variety of hiking trails ranging in difficulty from easy to strenuous. The hot, arid climate in Nevada combined with the rugged terrain makes hiking in the area particularly dangerous.

Overview

As park ranger, public safety is your primary job. In an effort to educate your visitors as to what difficulty they'll be confronting on the hiking trail, your goal is to deploy a web application on several kiosk computers in the visitor center. In addition to the basic trail routes and their difficulty ratings, your map will deliver more. For one thing, showing the terrain in highly visual relief (not possible with standard basemaps) will highlight a sense of the topography. Also, the map will reveal additional relevant information via pop-ups about each trail with a click. And with those clicks, the app will also display a graphic profile that shows the elevation gains and losses for the entire length of the trail.

This application can be built in a short time using Landscape Layers from the ArcGIS Living Atlas. No special skills are required.

▸ Build skills in these areas:
- Adding data to a map
- Adding raster data from Living Atlas
- Combining raster layers artfully using transparency
- Symbolizing lines and features
- Configuring a profile elevation web app
- Deploying a web app

▸ What you need:
- Publisher or administrator role in an ArcGIS organization
- Estimated time: 1 hour

Start Lesson

Esri.com/ArcGISBook/Chapter4_Lesson

The Importance of Where
How spatial analysis leads to insight

Spatial analysis allows you to solve complex problems and better understand where and what is occurring in your world. It goes beyond mapping alone to let you study the characteristics of places and the relationships between them. If the spatial component is important to the problem, spatial analysis lends perspective to your decision-making.

Spatial problem solving

Have you ever looked at a map of crime in your city and tried to figure out what areas have high crime rates? Have you explored other types of information, like school locations, parks, and demographics to try to determine the best location to buy a new home? Whenever we look at a map, we inherently start turning that map into information by analyzing its contents—finding patterns, assessing trends, or making decisions. This process is called "spatial analysis," and it's what our eyes and minds do naturally whenever we look at a map.

Spatial analysis is the most intriguing and remarkable aspect of GIS. Using spatial analysis, you can combine information from many independent sources and derive new sets of information (results) by applying a sophisticated set of spatial operators. This comprehensive collection of spatial analysis tools extends your reach toward answers to your questions. Statistical analysis can determine if the patterns that you see are significant. You can analyze various layers to calculate the suitability of a place for a particular activity. By employing image analysis, you can detect change over time. These tools and many others, which are part of ArcGIS, enable you to address critically important questions and decisions that are beyond the scope of simple visual analysis. Here are some of the foundational spatial analyses and the ArcGIS tools that get them done.

Understand Places

Analysis Tools: *Attribute queries, spatial queries; proximity analysis*

Detect Patterns

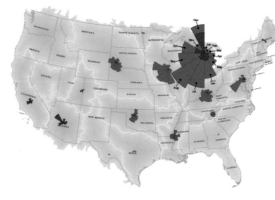

Analysis Tools:
Density analysis and cluster analysis

Find Locations

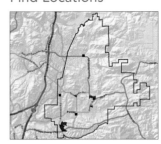

Analysis Tools: *Site suitability, location-allocation, cost corridors*

Determine Relationships

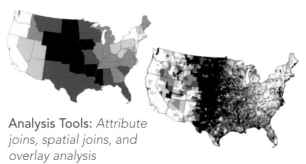

Analysis Tools: *Attribute joins, spatial joins, and overlay analysis*

Make Predictions

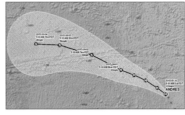

Analysis Tools: *Interpolation, regression, surface analysis*

Ask, calculate, interpret, decide, communicate

Spatial analysis is a process that starts by asking a question. Getting the question right is the key to working out what tools to use and what analysis to employ toward deriving a meaningful answer. Being able to better understand your data should help you to make appropriate decisions about your analysis—and anticipate how your choices will affect the results.

The web is now a source of vast amounts of data, and spatial analysis offers the means by which this data becomes valuable. As the value of analysis and spatial data gains adoption, methods and models for how to use analysis are becoming more widely available. Academics, professionals, communities, and individuals are all sharing their analytical techniques. These spatial data analysis practices cut across many disciplines. The applications are endless, and embracing a multidisciplinary approach to spatial analysis can bring significant rewards.

GIS analysis helps you to make informed decisions, but it doesn't make the decisions for you. Doing that requires your expertise. For example, you learn that multiscale web maps can require analysis to be done at multiple scales since, for many aggregated datasets, the results are only relevant for the scale at which the analysis is performed. You'll also need to learn to interpret results carefully—you should have an idea of the expected results and carefully check the analysis if the conclusions differ significantly.

As more communities see the power that analyzing their data with location brings, GIS technology is being pushed to new levels. Data from disparate sources is being combined, and new information is revealing new patterns and insights. Even the concept of geographic space is undergoing a challenge, as we map data using both physical and social geographies. Web maps bring spatial data and the notion of spatial analysis to everyone. The idea that location matters is no longer just the geographer's doctrine; its value has been widely recognized and embraced. Geography matters.

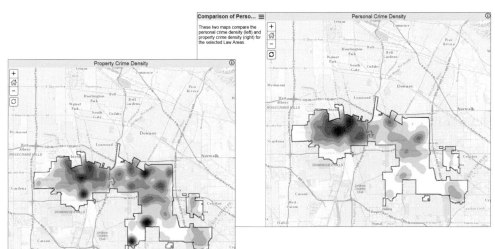

How is spatial analysis used?

Ask questions, derive answers

Spatial analysis is used by people around the world to derive new information and make informed decisions. The organizations that use spatial analysis in their work are wide-ranging—local and state governments, national agencies, businesses of all stripes, utility companies, colleges and universities, NGOs—the list goes on. Here are just a few examples.

Transportation
The Portland Bureau of Transportation uses spatial analysis to reveal the pattern of accidents and their relationship to traffic corridors.

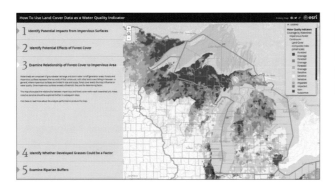

Water quality
NOAA developed an analysis methodology that uses land cover as an indicator of water quality.

Public health
This series of maps shows mosquito hotspots week by week. The information helps East Flagler Mosquito Control District in Florida target its suppression efforts.

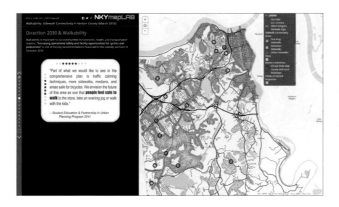

Urban planning

Kenton County, Kentucky, uses spatial analysis to map walkability in various communities and to identify gaps in sidewalk connectivity.

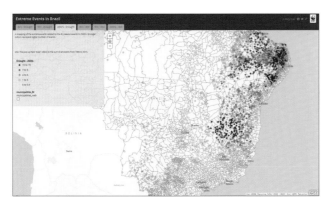

Conservation

WWF mapped extreme weather events, rainfall, and drought in Brazil over three decades. The trends and patterns help identify areas for future conservation projects.

Agriculture

This app from World Resources Institute lets government agencies and private companies find sites that are suitable for sustainable palm oil production in Indonesia.

Spatial data and spatial analysis

Most data and measurements can be associated with locations and, therefore, can be placed on the map. Using spatial data you know both what is present and where it is. The real world can be represented as discrete data, stored by its exact geographic location (called "feature data"), or continuous data represented by regular grids (called "raster data"). Of course, the nature of what you're analyzing influences how it is best represented. The natural environment (elevation, temperature, precipitation) is often represented using raster grids; whereas, the built environment (roads, buildings) and administrative data (countries, census areas) tend to be represented as vector data. Further information that describes what is at each location can be attached; these are often referred to as "attributes."

In GIS each dataset is managed as a layer and can be graphically combined using analytical operators. By combining layers using operators and displays, GIS enables you to work with these layers to explore questions and find answers.

In addition to locational and attribute information, spatial data inherently contains geometric and topological properties. Geometric properties include position and measurements, such as length, direction, area, and volume. Topological properties represent spatial relationships such as connectivity, inclusion, and adjacency. Using these spatial properties, you can ask even more types of questions of your data and gain new insights.

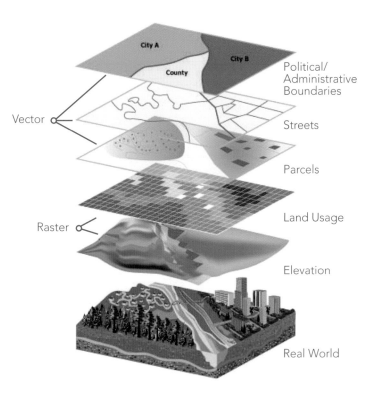

The idea of stacking layers containing different kinds of data and comparing them to each other based on where things are located is the foundational concept of spatial analysis. The layers are interlocking in the sense that they are all locked to true geographic space.

Visualization
What can my map show me?

In many cases, just by making a map you are doing analysis. That's because you're making the map for a reason. You have a question you want the map to help answer: Where has disease ravaged trees? Which communities are in the path of a wildfire? Where are areas of high crime? It's also because when you make a map, as with any analysis, you're making decisions about which information to include and how to present that information. Effective visualization is valuable for communicating results and messages clearly in an engaging way. Here are three key decisions that affect the information a map presents and the story it tells.

1. Choose scale

The scale of the map itself (the area you're showing) and the scale of the data you use both affect what your map will show. A classic example of how your choice determines the question answered is whether to show presidential election results by state or by county. While the state-level data does show a distinct national pattern, the county-level map reveals much more nuanced local and regional patterns. Map A answers the question, *What is the pattern of states (and electoral votes) won by each candidate?* Map B, about voting by county, better answers the question, *What is the distribution of Republican and Democratic voters in this election?*

Of course the area you're analyzing—your town, county, region, or state, for example—often determines the scale of data you use. But even at the city or county level, you might be choosing between mapping information using census tracts, block groups, blocks, or even lots.

Map A

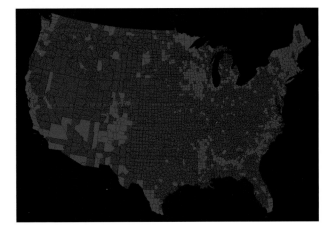

Map B

Visualization

2. Select style and attributes

The way you display the features on your map, the colors, line widths, and so on, is another decision that, while seemingly straightforward, helps clarify (or obfuscate) the information your map conveys. These two maps show the same location and the same features. Map C shows streets and highways in red, and streams and lakes in light blue. Map D shows streets and highways in gray, and streams and lakes in dark blue. Is the area highly developed or relatively pristine?

The same holds true in how you label the features on your map and the attributes you display in tables, pop-ups, and charts. Each selection is an opportunity for you to emphasize the information you need from the map and communicate that information to others (Map E).

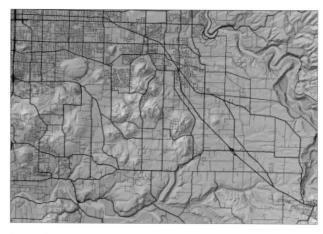

Map C

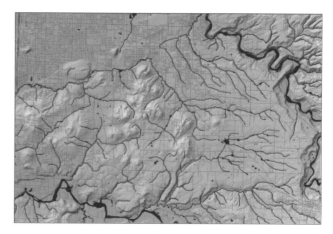

Map D

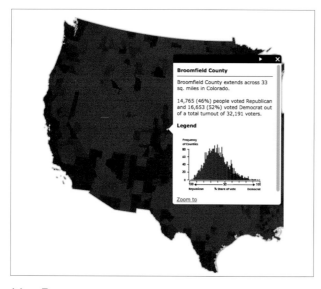

Map E

3. Determine classification scheme

Phenomena with similar values can be grouped together into classes so that similar areas can be seen clearly on the map. Visually, this can send an immediate message and allow you to gain a better understanding. The classification scheme you use defines which features fall into each group and, therefore, how the map appears. Once again, selecting an appropriate scheme is based on your data and the question you're asking. New smart mapping can help you make a suitable choice.

ArcGIS shows you the distribution of values and suggests an appropriate classification. But you can override this and choose from a number of standard classifications, or create your own scheme. These maps show the percentage of senior citizens in each census tract (dark red means a higher percentage). Map F displays the percentages using a continuous color scale, while map G uses five classes of equal range (known as "equal interval"). How much variation is there in the distribution of seniors across the county? From map F it looks like quite a few tracts have a high percentage of seniors while others have a very low percentage. Map G seems to indicate that most tracts have about the same percentage of seniors.

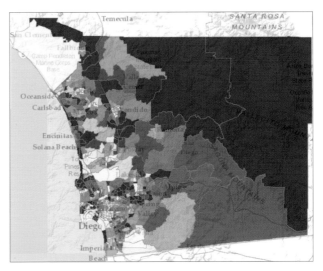

Map F

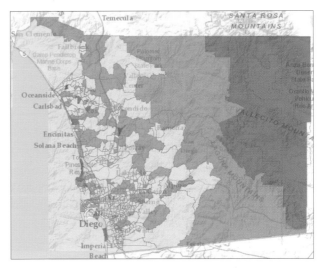

Map G

Exploring
What can my data tell me?

One of the most common types of spatial analysis you'll do in your new endeavor is exploratory; you'll explore your data using the myriad of analysis tools available in ArcGIS. Exploratory methods summarize map patterns and relationships and help you see what is happening where. Visualizing data distributions can reveal new information that can be effectively communicated, especially in conjunction with supplementary information such as pop-ups or graphs.

Descriptive statistics

Descriptive statistics help describe the main features of a dataset. They can be used to understand your data as part of your analysis or to quantify your data in some way. Combined with location, the main features of an area can be explained and mapped to communicate a clear message. Understanding what is happening is important.Additionally, knowing where it is happening is more powerful for interpreting and reporting results as well as for further inquiry.

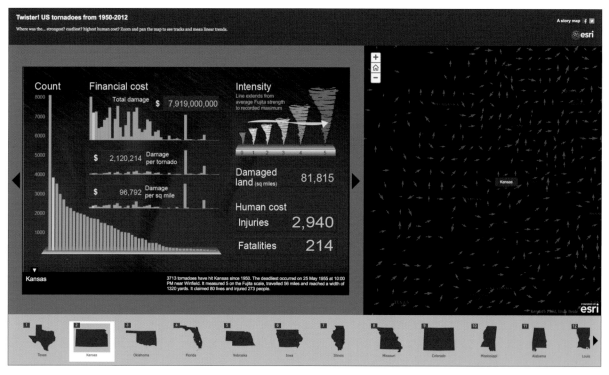

This app displays tornado tracks and uses descriptive statistics to summarize the human and financial costs of tornadoes over a six-decade period for each state.

Queries

Most analysis, whether spatial or not, starts by asking questions of the data (querying). Attribute queries use the values in the data, whereas, spatial queries use the location of the data. Querying data is an important process and in some cases, when used in sequence, can be used to solve a problem. For example, you can find suitable locations that meet a number of criteria, such as having a defined land type, elevation, and average temperature. In a GIS, queries create a temporary subset of the data that is shown on the map. You can then use this data in subsequent analysis, such as using descriptive statistics to explain the main characteristics of those particular features.

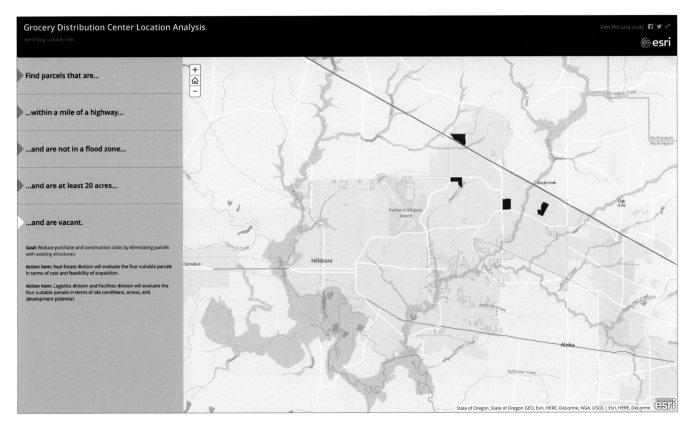

Attribute and spatial queries were used to find the four suitable parcels for a grocery distribution center—out of more than 17,000 parcels in the city.

Spatial relationships

A great deal of spatial analysis is concerned with the geographical location of a phenomenon in relation to another similar phenomenon or compared to other phenomena. You can learn a lot by exploring relationships between features within layers or even between different layers.

One type of spatial relationship you're probably familiar with is proximity. Proximity, or distance, allows us to find what's nearby. Although distance is often expressed in units of length, time and cost are examples of other measures that can be used to characterize the distance between phenomena.

While proximity analysis shows what things are near other things, overlay analysis shows you what things occur at the same location. This type of analysis is used to discover relationships between phenomena, such as the possible connection between the occurrence of a certain type of vegetation and environmental factors like elevation, soil type, and rainfall. Overlay analysis is also used to identify areas that meet a set of criteria you define in order to take some action, such as locating a new housing development or defining a protected habitat area.

Spatial patterns

Spatial patterns deal with the distributions of values (attributes) and the spatial arrangement of the locations. A continuous surface that shows the intensity of features or values based on sample observations can highlight areas of concentrations. By comparing the intensity of one type of event relative to another (such as a control group), you can discover potentially meaningful differences. Spatial clustering may indicate patterns of underlying processes, as similar processes often follow similar locating patterns. Areas of local concentrations of high or low values within a dataset can be identified and patterns can be more clearly visualized.

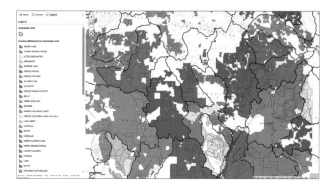

Overlay analysis was used to create this map showing grazing allotments in specific watersheds. State biologists will use the information to monitor water quality.

The distribution pattern of two species of mussels, zebra and quagga, can be clearly seen on this map. The invasive mussels pose a risk to native species as well as to the water distribution infrastructure.

Thought Leader: Linda Beale

The challenge is making complex data understandable

Dr. Linda Beale is a geoanalyst and expert in spatial epidemiology—the examination of disease and its geographic variations. Her work contributes to the development of Esri's ArcGIS analysis and geoprocessing software, and as a Research Fellow in Health and GIS at the Imperial College London, she led the effort to publish The Environment and Health Atlas for England and Wales.

Geography plays a crucial role in the area of health analysis. Fundamentally, it represents the context in which health risks occur; environmental hazards, risks, susceptibility, and health outcomes all vary spatially. Access to health care is characterized by both human and physical geographies. Furthermore, management and policy differ by location and resources are allocated geographically. Health is something that is important to everyone, but health analysis is challenging and demands a number of skills including epidemiology, statistics, and geographical information science. Spatial epidemiology is truly multidisciplinary and although complex techniques are required for analysis, results must be accessible to everyone.

All of these challenges were faced during the development of *The Environment and Health Atlas for England and Wales*. The atlas was developed with the ambitious goal of providing a resource—

for the public, for researchers, and anyone working in public health—with a collection of multiscale, interactive web maps that illustrate geographical distributions of disease risk and environmental agents at neighborhood scale.

Environmental monitoring and heath surveillance has advanced in recent decades, but emergencies continue to cause economic and social damage and, of course, loss of life. As the world becomes ever more interconnected both socially and economically, environmental and health impacts are felt at a wider scale than ever before. For example, following volcanic eruptions and nuclear accidents, or as a result of disease outbreaks such as avian influenza and Ebola, too often the impacts of environmental hazards fall disproportionately on the most vulnerable populations.

GIS offers the technology to explore, manipulate, analyze, and model data from multiple sources. With spatial analysis hazard mapping and predictions developed for risk assessment, you can use models to evaluate response strategies, and maps to illustrate preventative strategies and for risk communication and negotiation.

As technology has evolved, so have the science, the data, and the tools to test hypotheses and gain deeper insights into public health. We find ourselves at a time when, for many analyses, we are no longer awaiting technological or data advances. Instead, we should challenge ourselves to improve our understanding and public health through analysis.

Modeling
What can patterns tell you?

Some problems demand you go beyond exploring the data into quantifying relationships or formally testing hypotheses. This is where modeling comes in. Spatial modeling allows you to derive new data from values of existing data layers and to predict what might happen and where. Modeling often takes you into the realm of developing specialized workflows through programming. Creating scripts and automated workflows lets you efficiently query and process large amounts of data and implement more complex algorithms. Increasingly, the value of sharing methods and code through the web allows you to create complex workflows without the need to develop all the components. Knowledge is being shared by putting the real power of spatial analysis into the hands of more people.

Modeling processes

With an understanding of the processes at work in the natural or human environment, additional features can be modelled from spatial data. Using an elevation surface, for example, you can derive information and identify features that were not readily apparent in the original surface, such as contours, angle of slope, steepest downslope direction (aspect), shaded relief (hillshade), and visible areas (viewsheds). You can model the flow of water across Earth's surface, deriving runoff characteristics, understanding drainage systems, and creating watersheds. For certain types of analysis, like creating wildlife suitability surfaces, multiple layers of information can be brought to bear on the problem. The plight of the mountain lion has been carefully studied in the analysis documented below, which incorporates elevation, terrain ruggedness, vegetation cover, protected land status, and other factors to help planners build wildlife corridors for the species' survival.

This GIS app shows how GIS was used to model areas that cougars would be likely to traverse through the mountains and wildlands near Los Angeles. Wildlife conservation experts stress the need to identify safe corridors, including natural bridges, so the big cats in isolated populations can find each other.

Interpolating values

Rarely is it possible to gather data for all locations continuously; there is invariably a spatial or temporal scale beyond which the data is unknown. GIS allows you to predict values at unsampled sites from measurements made at point locations within the same area. You can choose from a number of different interpolation methods, depending on your data and the phenomena you're modeling. Geostatistical methods, for example, provide interpolated values and, additionally, measures of uncertainty for those locations. The measurement of uncertainty is critical to informed decision-making, providing information on the possible outcomes (values) at each location.

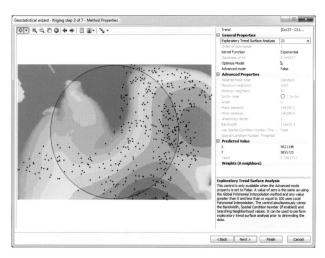

Kriging is used to predict cesium-137 soil contamination levels for the entire area surrounding Chernobyl, from a set of soil samples taken at testing locations.

Modeling spatial interaction

Spatial interaction models are used to estimate the flow of people, material, or information between locations. These models consider the attractiveness of a place (that is, some measure of why you want to go there) and the cost of travel (either financial or time cost) to get there. The measures are used to calculate the proportion of demand at each place.

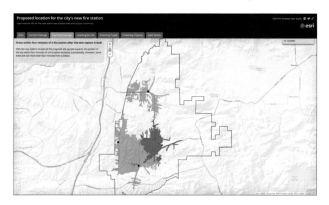

A spatial interaction model identifies the best site for a new fire station. The analysis assures that the selected site, along with existing stations, conforms to the pattern of each fire station being within a four-minute travel time to homes in its immediate coverage zone.

Quickstart

Discover new insights, knowledge, and understanding with spatial analysis

Spatial analysis cuts across several key aspects of the ArcGIS system. These Quickstart resources are intended to get you oriented.

▸ **Spatial analysis online**
The analytic capabilities of the online part of the system are accessed through the Analysis button on the ArcGIS map viewer:

License: For you to perform analysis online, the administrator of your organization needs to grant you privileges to create content, publish hosted feature services, and do spatial analysis.

▸ **Join the Learn ArcGIS organization**
If you're not already a member of an ArcGIS Online organization, you can join the Learn ArcGIS organization designed for students (see page 15). This will allow you to use all the online analysis tools and complete the Learn ArcGIS Lesson on page 78.

▸ **Install the ArcGIS Trial**
You can also install a 60-day Trial for the entire ArcGIS platform which includes 60 days worth of ArcMap and ArcGIS Pro for the lessons in chapters 2 and 6.

▸ **Online case studies**
An impressive set of spatial analysis case studies are on the ArcGIS Analytics website.

▸ **Spatial Analysis MOOC**
This Massive Open Online Course (MOOC) runs periodically. Spatial analysis focuses on location to gain a deeper understanding of data. Spatial analysis skills are in high demand by organizations around the world. You'll get free access to the full analytical capabilities of ArcGIS Online, Esri's cloud-based GIS platform.

▸ **Spatial Analyst**

Watch a video introduction to the Spatial Analysis MOOC
Esri.com/ArcGIS Book/Chapter5_Video

ArcGIS for Desktop
ArcGIS Spatial Analyst is an extension of ArcMap that augments the capabilities of ArcGIS for Desktop by adding a range of raster spatial modeling and analysis tools. It is used to solve complex problems such as optimally locating new retail stores or determining the most promising areas for wildlife conservation efforts. While beyond the scope of this book, it's an important tool in the serious analyst's kit.

ArcGIS Pro
The Spatial Analyst extension for ArcGIS Pro provides a similar suite of tools for that application. Pro is especially good for 3D and other data intensive analyses like terrain modeling, surface interpolation, suitability modeling, hydrological analysis, statistical analysis, and image classification.

Learn ArcGIS membership does not include a License for ArcMap or Pro; if you don't have ArcGIS for Desktop, you'll need to activate the free ArcGIS Trial (see page 15).

Learn ArcGIS Lesson

Evaluate, analyze, plan

▸ **Overview**

As an analyst for an Oregon-based real estate development company, you help your company find suitable areas for mixed-use developments. These developments have shops and restaurants at ground level and three or four floors of housing above. The target demographic group for both the retail businesses and the housing is people in their twenties and thirties who live nearby. You know from experience that when young adults are already clustered in an area, housing units are easier to rent, and ground-level businesses are more likely to thrive.

Start Lesson

Esri.com/ArcGISBook/Chapter5_Lesson

In this project, you'll evaluate areas in Gresham, Oregon, a city of about 100,000 people located east of Portland. You'll look for suitably-zoned areas where young adults are known to live and where renting is common. Since many of your prospective tenants will commute to Portland, you'll also look for areas that afford easy access to the city's light-rail system. While these are not the only factors that determine a good site, they are important criteria that will help the company refine its search.

▸ **Build skills in these areas**
- Enriching layers with on-demand attributes
- Filtering layers with logical queries
- Adding and calculating table fields
- Deriving new locations through analysis
- Creating time-based accessibility zones

▸ **What you need**
- Membership ArcGIS Online organization
- Estimated time: 60 minutes

Mapping the Third Dimension
A change in perspective

3D is how we see the world. With 3D Web GIS, you bring an extra dimension into the picture. See your data in its true perspective in remarkable photorealistic detail, or use 3D symbols to communicate quantitative data in imaginative ways, creating better understanding and bringing visual insight to tricky problems.

The evolution of 3D mapping

Throughout history, geographic information has been authored and presented in the form of two-dimensional maps on the best available flat surface of the era—scrawled in the dirt, on animal skins and cave walls, hand-drawn on parchment, then onto mechanically printed paper, and finally onto computer screens in all their current shapes and sizes. Regardless of the delivery system, the result has been a consistently flat representation of the world. These 2D maps were (and still are) quite useful for many purposes, such as finding your way in an unfamiliar city or determining legal boundaries, but they're restricted by their top-down view of the world.

Three-dimensional depictions of geographic data have been around for centuries. Artistic bird's-eye views found popularity as a way to map cities and small-extent landscapes that regular people could intuitively understand. But because these were static and could not be used directly for measurement or analysis, they were often considered mere confections, or novelties, by serious cartographers, not a means of delivering authoritative content.

However, this is no longer the case since ArcGIS introduced the concept of a "scene," which is actually more than just a 3D view. In a scene you can also control things like lighting, camera tilt, and angle of view. This means that the mapmaker can craft a scene that creates a highly realistic representation of geographic information in three dimensions, which provides an entirely new way for the audience to interact with geographic content. Spatial information that is inherently 3D, like the topography of the landscape, the built world, and even subsurface geology, can now be displayed in a way that is not only intuitive and visual but also quantitative and measurable, meaning we can do real analysis and hard science with 3D data.

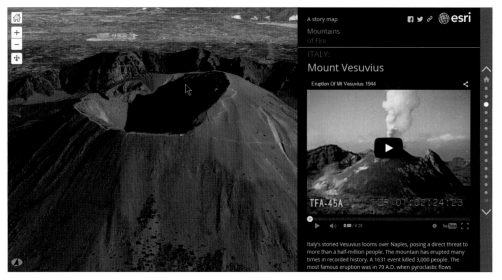

Some stories lend themselves very well to 3D storytelling. The Mountains of Fire story map is composed of a number of 3D web scenes.

Advantages of 3D

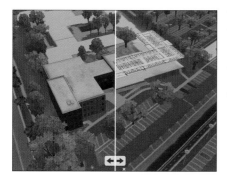

Vertical information

The most obvious advantage of a scene is its ability to incorporate vertical (and thus *volumetric*) information—the surface elevation of mountains, the surrounding landscape, the shapes of buildings, the flight paths of jetliners. It's the power of the z-axis.

This web scene was made prior to the construction of the northern addition to Building M on the Esri Redlands campus. Rather than create a detailed, 3D interior of the (still changing) floor plan, the plans were instead geo-referenced in three dimensions to provide quick, interactive feedback.

Intuitive symbology

In 3D, the extra dimension enables you to include more readily recognized symbols and, therefore, make your maps more intuitive. You are able to see all the "data" from all viewpoints in situ. Every symbol that you recognize on a map saves you the effort of referring to the legend to make sure you understand what it shows.

No need for a legend on this 3D scene. The elements like the soccer field, the palm trees, and the structures are all instantly recognizable.

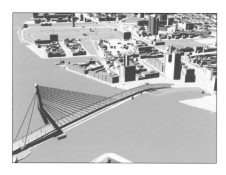

Showing real-world bird's-eye views

Many of man's earliest maps, particularly of cities and smaller human habitations, were portrayed as scenes. These stylized maps were created as static 3D bird's-eye views and were very successful in providing understanding of a place. Today's GIS authors interact and see these views from many perspectives.

Fly like a bird through this 3D scene of the City of Rotterdam, Netherlands.

Human-style navigation

For most of our living moments, we experience the world within 10 feet of the ground. 3D allows us to replicate this view. By presenting data from this approachable perspective, the size and relative positions of objects are intuitively understood by the viewer wandering virtually through the scene. There's no need to explain that you're in a forest or that a lake is blocking the route—the perspective immediately makes the features recognizable.

Important 3D terminology
Getting the z-terminology straight

Maps and scenes

GIS content can be displayed in 2D or 3D views, and there are a lot of similarities between the two modes. For example, both contain GIS layers, both have spatial references, and both support GIS operations such as selection, analysis, and editing.

However, there are also many differences. At the layer level, telephone poles might be shown in 2D as brown circles, while the same content in 3D could be shown as volumetric models—complete with cross members and even wires—that have been sized and rotated into place. At the scene level, there are properties that wouldn't make sense in 2D, such as the need for a ground surface mesh, the existence of an illumination source, and atmospheric effects such as fog.

In ArcGIS, we refer to 2D as "maps" and 3D as "scenes."

Local and global

3D content can be displayed within two different scene environments—a global world and a local (or plane) world. Global views are currently the more prevalent view type, where 3D content is displayed in a global coordinate system shown in the form of a sphere. A global canvas is well suited for data that extends across large distances and where curvature of the earth must be accounted for, like global airline traffic paths or shipping lanes.

Local views are like self-contained fishtanks, where scenes have a fixed extent in which you work. They are better suited for small-extent data, such as a college campus or a mine site, and bring the additional benefit of supporting display in projected coordinate systems. Local views can also be very effective for scientific data display, where the relative size of features is a more important display requirement than the physical location of the content on a spheroid.

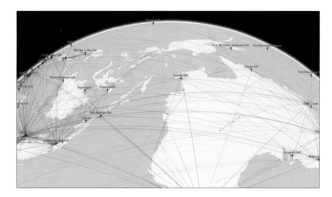

This scene shows the global pattern of airports and the interconnected air traffic routes using the openflights.org data.

In this local scene, Dealey Plaza in Dallas serves as the backdrop for a 3D recreation of the events that shocked the world on November 22, 1963.

Surfaces

A surface is like a piece of skin pulled tight against the earth. Surface data by definition includes an x, y, and z value for any point on it. A surface can be a physical thing that exists in the real world, like a mountain range, or it can be an imagined surface that might exist in the future, such as a road grading plan. It can even show a theme that only exists conceptually, such as a population density surface. Surfaces come in a wide variety of accuracies, with anything from high-resolution, 1-inch accuracy all the way down to a low-resolution surface with 90 meters or coarser accuracy.

Surfaces are fundamental building blocks for nearly every scene you will create because they provide a foundation upon which other content can be draped. Sometimes the surface itself is the star of the show (like a scene of Mt. Everest). Other times the surface serves a more humble role of accommodating other crucial scene data, such as aerial imagery or administrative boundaries. And surfaces can also provide base-height information for 3D vector symbols, like trees, buildings, and fire hydrants, that might otherwise not "know" their vertical position within the scene.

This scene presents surfaces of interesting places on our planet, featuring the World Imagery basemap along with Terrain 3D layers. You can click on the slides in the scene to explore them and navigate the scene to see different perspectives for each place.

Real size and screen size

Symbolizing features using a real-world size is extremely common in 3D. For example, it's expected that buildings, trees, and light poles would all be displayed at the same relative size in the virtual world as they exist in reality. Even some thematic symbols, like a sphere showing the estimated illumination distance of one of the light poles, will help communicate the notion of a real-world size.

However, it is also useful to have symbols in the scene that use an on-screen size instead. That is, as you zoom in and out within the view, the symbol always displays with the same number of pixels on the screen. This effect is analogous to a 2D map layer whose symbol sizes do not change as you move between map scales.

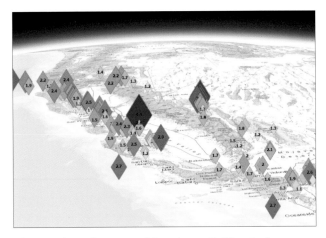

This earthquake map of Southern California features screen-size symbols that remain the same size regardless of how far or where you zoom in and out.

Representing the world in 3D

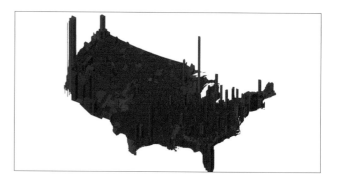

Photorealistic

Photorealistic views are essentially attempts to recreate reality by using photos to texture your features. These are by far the most common type of scene, with enormous amounts of effort put into making the virtual world look exactly like you were there in person. Authors of this content are making virtual worlds for simulation, for planning and design, and for promotional videos. The specification remains very simple: *Look out the window and make the virtual world appear like that.*

In a GIS context, photorealistic views are extremely well suited for showing the public how a place has changed, or is expected to change, through time. That could mean what the cityscape will look like after a proposed building has been constructed, or what a region looked like when dinosaurs roamed the earth. A photorealistic view takes the onus from users of imagining what the state of the world would look like, and simply shows them.

3D cartographic

For adventuresome cartographers, using 3D elements to represent data and other non-photorealstic information is the next frontier. The idea is to take 2D thematic mapping techniques and move them into 3D. These maps are powerful, eye-catching, and immersive information products, often viewed as navigable scenes or packaged as video to control the user's experience and deliver maximum impact.

Virtual reality

A 3D view quickly starts to feel like virtual reality when photorealistic and thematic techniques are used in combination. The photorealistic parts of the scene provide a sense of familiarity to the user, and the thematic parts can convey key information. Slip on an Oculus headset and you're suddenly immersed in a 3D world.

What makes a great scene?

Look and feel

By intent, 3D views are designed to be immersive. We experience and see spaces in 3D. People viewing the content are, effectively, invited to imagine themselves within the scene as they fly around. This means that the styling, or the look, of the world surrounding them can have a strong impact on how they feel about the scene in general.

For example, a city shown with dark lighting and heavy fog lends a sense of foreboding or decay to the area, while a bright and sunny depiction of the same city, with people and cars, implies that the city is vibrant and safe—think Gotham versus Pleasantville.

Styling 3D content

The styling of the GIS content itself, within the scene, also has a big impact on the look and feel of the scene. There are basically three choices available to you: fully photorealistic, fully thematic, or a combination of photorealistic and thematic.

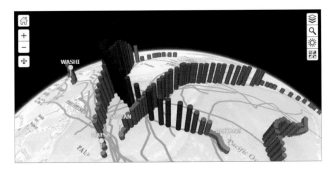

In 2005, there were 23 typhoons that made their way through the Western Pacific Ocean region. This global-scale scene uses thematic vertical columns to describe their path and relative windspeeds, while pop-ups provide access to associated satellite photographs.

Thematic

Thematic views model and classify reality in a way that communicates spatial information more effectively. Thematic 3D views use common 2D cartographic techniques, such as classifications, color schemes, and relative symbol size, to simplify the real world into something that can be more readily understood. 3D scene authors create schematic, simplified representations to more effectively convey some key piece of information, particularly for scientific visualization.

For GIS users, thematic content can be an effective, and eye-catching, way to display more than just *where* something is—it can also show key *properties* about that thing. As in the example below, typhoon data points can be symbolized to show both the path of the storm and its changing windspeed.

In this web scene, the colors represent the number of key viewpoints (0-3) that each section (or "panel") of a proposed building can see.

Thought Leader: Nathan Shephard
The rise of the 3D cartographic scene

When people talk about seeing an amazing computer-generated 3D view, they are nearly always talking about a realistically rendered view. You know, the one with ray-tracing and ambient lighting and reflective surfaces, where it looks so much like the actual world you can almost touch it. While this type of view is useful for conveying certain types of geographic information—such as a proposed future cityscape—it is not the right way to render everything. That is, in the same way that every map is not an aerial image, every 3D view should not be an attempt to recreate the real world.

GIS users share maps and scenes with one common goal—to communicate spatial information—and careful use of thematic symbols in 3D can be as effective, or even more effective, than similar techniques in 2D. For example, showing tree features as colored spheres on sticks (with red representing those that need to be trimmed) is much more to the point than displaying those same trees as highly realistic models covered with leaves and branches. The size of the spheres can still contain elements of the real world, such as the height and crown

Nathan Shephard is a technology evangelist and 3D GIS engineer at Esri, as well as an independent game developer.

width of each tree, but the real value of the symbols comes from their cartographic display—a simpler, more representative display that provides an immediate visual understanding of which trees are important. The advantage of using 3D is that a sphere on a stick still looks enough like a tree that you don't need to have an explicit legend saying *Tree*.

For centuries, cartographers have been limited to two dimensions. They've experimented with more effective ways of communicating spatial information through the clever use of symbols and classifications and colors. The existence of medieval, bird's-eye view maps shows that many grasped the power of the third dimension even if they didn't have the tools to fully explore them. But now, suddenly, everyone has those tools, and 3D cartographers have the extra, wonderful third dimension to work with.

 Video: How to author web scenes using ArcGIS Online
Esri.com/ArcGISBook/Chapter6_Video1

Who uses 3D cartography?

3D mapping and cartography have applications across a broad swath of industries and in government and academia. The examples featured here hint at the possibilities.

Take some time to click through these apps on your computer. These and many other innovative examples are collected in the ArcGIS Web Scenes gallery.

Emergency managers

The deadliest landslide event in US history occurred in Oso, Washington. This quickly assembled 3D comparison was used in the recovery efforts.

Historians

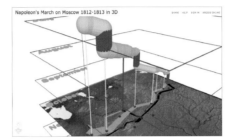

A 3D update to one of the most famous maps of all time brings new understanding to Napoleon's disastrous Russian campaign of 1812.

Tourism boards

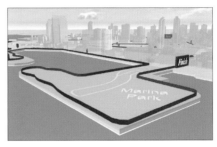

This walking map of San Diego uses a kind of augmented reality to overlay points of interest and paths over a stylized 3D city view.

Crime analysts

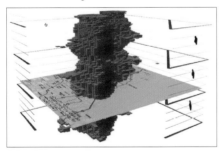

Massive datasets, like three years' worth of crimes committed in Chicago, lend themselves to 3D visualization. In this case, the z-axis is actually being used to depict time.

Urban planners

This 3D map of Philadelphia was created to show the impact of sunlight and visibility for a proposed high-rise development downtown.

Architects and facility managers

The mapping of building interiors as well as exteriors is an informative and immersive way for people who build and manage facilities to think about their assets.

Quickstart

Take your maps into the third dimension with these parts of the GIS system

▸ **The Scene Viewer**
The basic ArcGIS Scene Viewer allows you to work immediately in 3D space. It functions with desktop web browsers that support WebGL, a web technology standard built into most modern browsers for rendering 3D graphics. Check out this gallery of scenes to verify that your browser is properly configured.

▸ **3D in ArcGIS Pro**
ArcGIS Pro is a modern 64-bit desktop application that has extensive 3D capabilities built in. You can work with 2D and 3D views side by side. ArcGIS Pro is included in the free 60-day ArcGIS Trial.

▸ **Esri CityEngine**
CityEngine is an advanced tool for scenario-driven city design and developing rules for creating procedurally built data.

▸ **Terrain and basemap overlays**
Each scene starts with a basemap draped on the 3D elevation surface of the world. Zoom to your area of interest and begin to add your operational overlays.

▸ **What is the purpose of your scene?**
Before you start designing your new scene, you need to know its purpose. What is the message or information you intend to convey?

The answer to that question will help you design many elements of your scene.

● For example: Does curvature of the earth help or hinder the message (global view vs local view)?

● Will thematic styling distract from or augment the GIS information (photorealistic vs thematic layers)?

● Do users need to zoom in close to the ground (minimum surface resolution)?

● What basemap do users need draped on the ground for context (imagery, cartographic maps, thematic)?

The key point is that each of your decisions should be rooted in why you are building the scene in the first place.

Share Web Scenes using ArcGIS Pro

Esri.com/ArcGISBook/Chapter6_Video2

Learn ArcGIS Lesson

Create 2D and 3D maps to analyze flooding in Venice, Italy

Spanning a series of islands in a shallow lagoon, the city of Venice is renowned for its beauty. But that beauty comes at a cost. The lagoon's tidal patterns mix with the island's low elevation to cause *acqua alta* (high water), a periodic flooding that affects most of the city. Although not a threat to human life, *acqua alta* impedes transportation and endangers Venice's priceless architecture— and the problem is getting worse.

▸ Overview
In this project, you'll travel to Venice with ArcGIS Pro. You'll build a 2D map of the city with canals, structures, and some of Venice's most famous landmarks. Then, you'll convert the map into 3D.

You'll analyze and quantify the threat of *acqua alta*, before giving your scene a realistic appearance to show others.

▸ Build skills in these areas:
- Adding data to a map and editing features
- Creating a scene and analyzing raster data
- Applying 3D symbology

▸ What you need:
- ArcGIS Pro
- An ArcGIS Online organizational account with a Pro license enabled by your organization's administrator

Start Lesson

Esri.com/ArcGISBook/Chapter6_Lesson

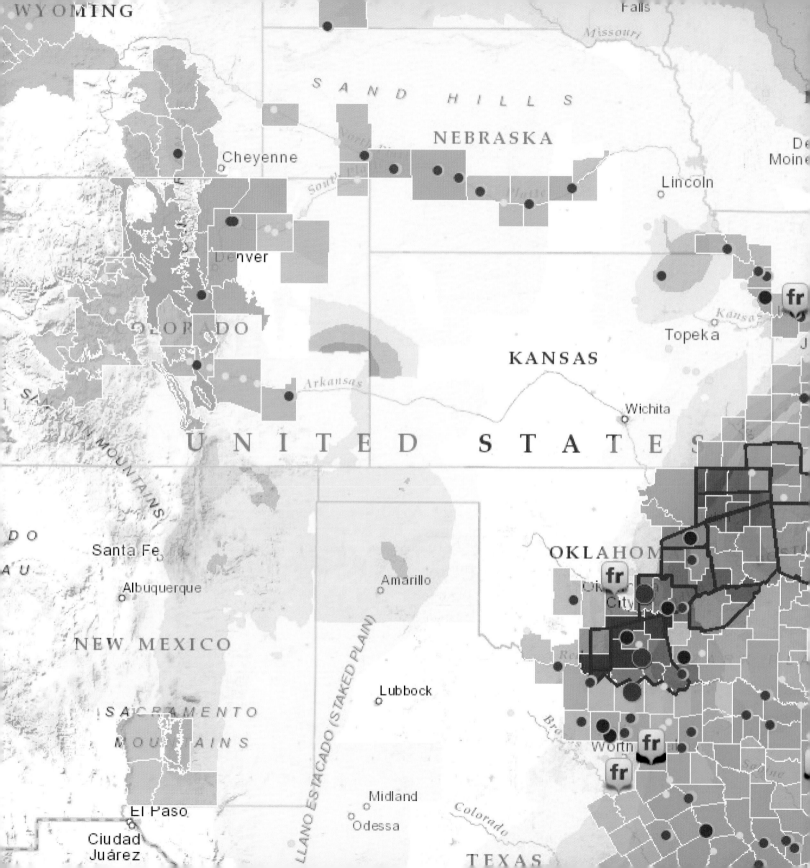

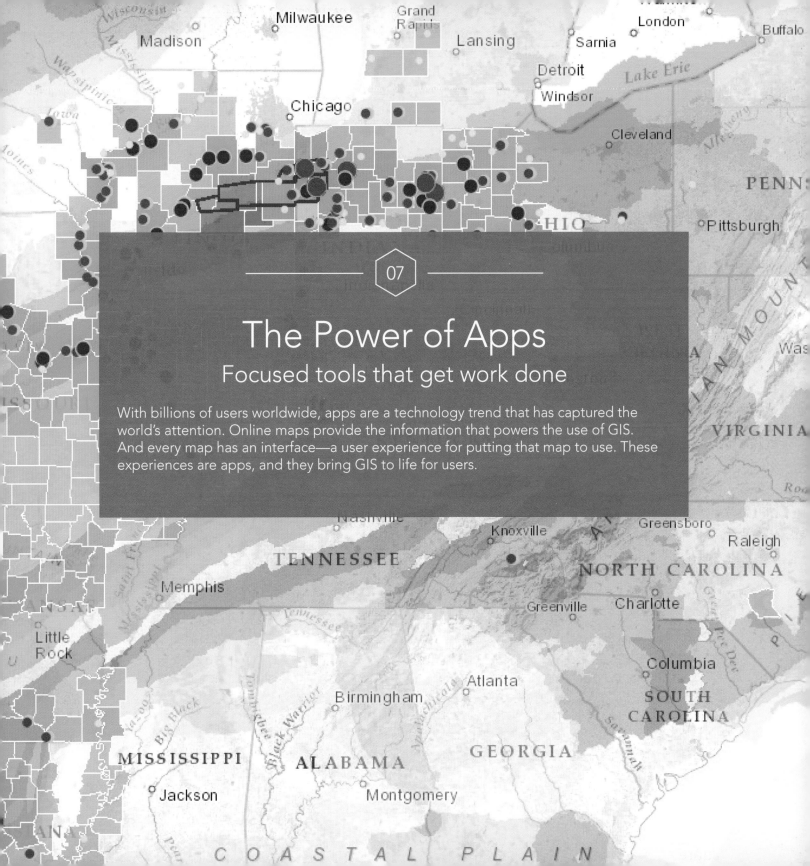

The Power of Apps
Focused tools that get work done

With billions of users worldwide, apps are a technology trend that has captured the world's attention. Online maps provide the information that powers the use of GIS. And every map has an interface—a user experience for putting that map to use. These experiences are apps, and they bring GIS to life for users.

The rise of spatially intelligent apps

Apps are lightweight computer programs designed to run on the web and on smartphones, tablets, and other mobile devices. GIS apps are a special breed; they're map-centric and spatially aware. Seemingly overnight, apps are ubiquitous. Billions of people worldwide run them in their web browsers, on computers and, of course, on their mobile devices. Creating interesting geographically aware apps is now within your reach. From the intuitive story map app and Web AppBuilder to the app collection for your smartphone and tablet, the technology required to deploy highly effective apps that can really engage an audience is accessible.

Apps are often built around targeted workflows that deliver streamlined user experiences. They're designed to guide users through specific tasks, to show just the data that is required for that task, and to allow ease of communication of your message. This chapter shows where apps come from and how you can start to create your own. It explores some of the innovative ways that apps are being used to do real work with ArcGIS. You'll discover ArcGIS apps for the work you do, no matter what the task or the device. Need to collect data in the field? There's an app. Need to share your data with the public? There's an app for that, too. Whether you're managing a mobile workforce, creating a geolocation startup, or looking for innovative ways to share your information in a useful and meaningful way, apps are your way to do that.

Thought Leader: Abhi Nemani
Shaking things up in the City of L.A.

The notion of government as a platform fundamentally means that innovation can come from anywhere—inside government or out. That's because the data that has typically been locked within government silos is now openly available for anyone to use, remix, and build upon. This only works if citizens are interested and willing to step up and get involved. And they are, particularly when there is an opportunity related to a place where they live and care about. That's why open geospatial data is so attractive to those wishing to make government more of a platform. People care about the buildings near them, the schools their kids are eligible for, and the new businesses on their block. They care even more about how all those things together form a vision of health and well-being for their community. GIS helps bring that vision to life.

I firmly believe the best way to connect with people is to go to where they are. Thus, I also believe the next step for open data is to go more proactively to where citizens are. That means not just posting the restaurant inspection scores on an open data portal, but integrating those scores with restaurant search engines such as Yelp. It means publishing building inspection records on Zillow. It means incorporating traffic and transit information on Google, Waze, and Apple. All these things are, in fact, already happening. Indeed, I think it's just the start. What's next is exciting: what are the other consumer apps that could be made more civic?

We should not only be creating beautiful and elegant citizen-facing solutions but also developing more effective tools for public servants

Abhi Nemani is a writer, speaker, organizer, technologist, and GIS power user. He serves as the first Chief Data Officer for the City of Los Angeles, where he leads the city's efforts to build an open and data-driven L.A.

to better serve the community: data analysis tools to prioritize service delivery, workflow systems to streamline communications, and data collection tools to speed up on-the-ground reporting, just to name a few. These are force multipliers. They enable public servants to serve more people better, faster, and smarter.

Adapted from *ArcNews:*
Read the entire interview

Case study: US Geological Survey

In 2009, the US Geological Survey began the release of a new generation of topographic maps (US Topo) in electronic form, and in 2011, complemented them with the release of high-resolution scans of historical topographic maps of the United States dating back to 1882. The topographic map remains an indispensable tool for everyday use in government, science, industry, land management planning, and recreation.

Historical maps are snapshots of the nation's physical and cultural features at a particular time. Maps of a particular area can show how the area looked before development and provide a detailed view of changes over time. Historical maps are often useful to scientists, historians, environmentalists, genealogists, citizens, and others researching a particular geographic location or area. The USGS, with help from Esri, has created an app that lets you view this extensive collection of USGS topographic maps in one central location. The USGS Historical Topographic Map Explorer makes it easier to dig into and enjoy this library of more than 178,000 historical maps in a web application that organizes the maps by space, time, and map scale.

Using the USGS Historical Topographic Map Explorer is easy—just follow the numbered steps at the left of the interface pane. The choices you make will update the adjacent map view.

Here's how to get the most out of the app:

- Find the area you want to explore
- Use the timeline to select the maps
- Compare the maps

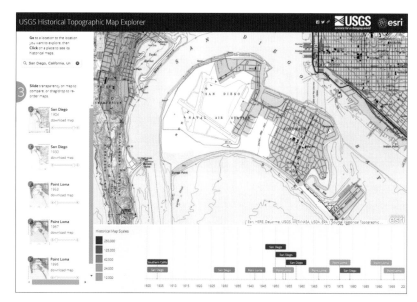

The USGS was created in1879, charged with the "classification of the public lands, and examination of the geological structure, mineral resources, and products of the national domain." This simple-to-use app compiles this impressive body of historical mapping work into an engaging user experience.

Where do apps come from?

You can get usable apps on your own devices and those of your audience from a number of sources. They range from the off-the-shelf apps from Esri and other developers in the business community—and "roll-your-own" apps that you configure and publish using templates and builders—to fully customized solutions built by developers with software development kits (SDKs) and application program interfaces (APIs).

Ready-to-Use ArcGIS apps
ArcGIS includes a suite of apps that are ready to go and free to use if you have an ArcGIS account. Mapping apps, like ArcGIS Explorer for the Mac, provide a way to manage a collection of data.

ArcGIS Explorer for the Mac.

ArcGIS Marketplace
ArcGIS Marketplace is where you can obtain not only apps and data services from Esri but also its distributors and partners. All the apps in the Marketplace are built to work with ArcGIS Online, so they can easily be shared with ArcGIS Online groups and users within your organization.

The Satellite Tasking and Archive app allows ArcGIS users to easily access and program Airbus Defence and Space satellites.

ArcGIS Solutions
ArcGIS Solutions are available for a range of industries such as local and state government, emergency management, utilities, telecommunications, military, and intelligence.

Web AppBuilder for ArcGIS
Web AppBuilder for ArcGIS enables you to create custom web mapping applications in an intuitive environment.

Visit developers. arcgis.com and https://github. com/Esri/ for developers resources, code, and templates.

Code your own
Coding custom apps requires the greatest level of effort, but provides the most flexibility.

Solve a problem with an app

The following pages describe almost a dozen different things that can be done with apps. As you read through them, think about what it is that you need to do that could fit into one of these patterns. We'll show you how to get started using these in the Quickstart section, so for now just consider the possibilities.

Tell a story
You can author a story map (the focus of chapter 3) fairly easily by choosing from the many storytelling narrative styles offered in ArcGIS to bring your data to life. Story map apps combine maps with rich narratives and multimedia content that connect with an audience and keep them engaged.

Here's the real story of the desperate effort to save Europe's artistic and cultural heritage.

Engage with users
Map-centric apps offer an exciting and engaging way to publish your geographic information. When you can interact with a live map, send queries back to the GIS of the world, have the app follow you and alert you when you get certain places, all of a sudden you've got something very powerful and engaging because the experience is personal and familiar to you.

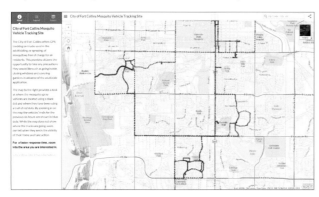

West Nile virus is a mosquito-borne, potentially fatal illness that has recieved considerable public attention. In response, the city of Fort Collins, Colorado, keeps its citizens abreast of their mosquito abatement operations in real time on the web.

iGeology is a free smartphone app that lets you take over 500 geological maps of Britain wherever you go to discover the landscape beneath your feet.

Collect data

Apps provide the interface for the efficient collection of spatial data. You can work offline in the field with a native data collection app or employ crowdsourcing or VGI (volunteered geographic information) using a web app. Chapter 8 delves deeper into this idea of the device as a sensor.

The dramatic rise of high-powered, smart mobile devices means that almost everyone has a highly capable sensor in their pocket. This opens mobile data collection to almost every organization or person in the developed world. Big strides in cellular and mobile infrastructure have allowed GIS apps to be deployed in even the poorest regions internationally. The ubiquitous presence of handheld devices is being leveraged as a data collection tool very successfully throughout the world.

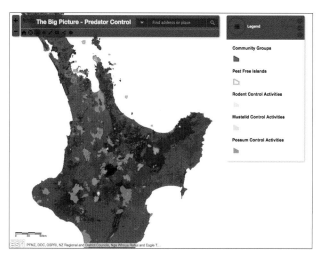

This interactive map of New Zealand shows recent and historical predator control activities and allows geographic information to be added directly by citizens capturing the collective wisdom of the "crowd."

MobileMap from forest industry leader Mason, Bruce & Girard provides GIS functionality to users in the field, so they can collect data and perform complex visualization and discovery.

Answer questions through analysis

Geographic insight is often the best way to answer pressing questions. Overlaying multiple layers on a map and analyzing them with advanced spatial models can highlight relationships that are not otherwise apparent. Knowledge workers using ArcGIS have the power to build models to answer almost any question, from where to locate a new facility to finding areas most at risk.

With the Geoplanner app, a city architect/analyst combines factors such as slope, aspect, population, and distance to water to understand the implications of many different development scenarios.

Open your data

Building open data portals provides public access to your authoritative data. Branded web apps allow you to share your data in ways that are easy to search and explore. Open Data lets you be part of a worldwide community of geodata content curators. To learn more, visit ArcGIS Open Data online http://opendata.arcgis.com/

The widely adopted app, ArcGIS Open Data, provides an experience that enables GIS organizations to readily share their data with their community.

Provide geo-alerts

Location may be front and center, meanwhile intelligent analysis runs invisibly in the background. Without distracting users from what they are trying to do, these background services allow alerts to be fired off when you need them. For example, when the device (and of course the user holding it) enters a certain area where hazardous materials are located, a "geo-fence" can warn them.

The Sensimob Field Team Messenger app (a native app for Android devices) combines GPS tracking functions with team-based social features allowing team members to keep track of each other's whereabouts in real time.

Track and monitor

Knowing where workers are and keeping them safe and informed may be the difference between profit and loss—or even life and death. Device location sensors can be plugged into apps to record critical data to your operations. Spatial data also can be gleaned from other data sources, such as Twitter feeds, to help show patterns.

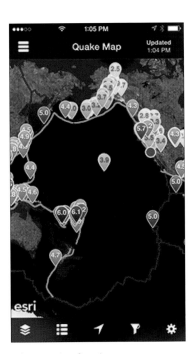

The quake feed app uses your location to send you alerts about earthquakes near you.

This live train app tracks the location of trains in Helsinki, Finland.

Get around

Identifying your location and generating routes and directions help to keep your users moving in the right direction (even onto private roads or inside buildings). With the rich geocoding and routing services provided by Esri or through your own organization's locators and networks, everyone can get around quickly and efficiently.

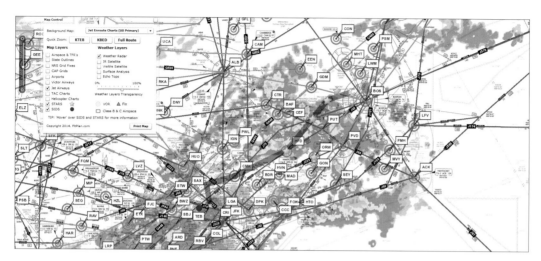

FltPlan Go is an Electronic Flight Bag (EFB) for pilots. It offers a variety of tools and features including moving maps, geo-referenced approach plates, airport information, airport diagrams, up-to-date fuel prices, in-flight weather, and more.

Manage operations

To control and plan operations you need to know where things are located and what their current conditions are. Pairing infographic displays with spatial location can be the most effective way to visualize an operation and communicate a management plan. Whether you are in the office or on the go, native and web apps help you make decisions.

Sometimes apps fill the entire wall of a room. Operations centers run ArcGIS dashboards on big displays.

Add demographic context to your maps

Esri Insights is a powerful web app that allows users to visualize demographic information about almost any area in the world. Gain deeper insight into the facts and dominant characteristics of your neighborhood or about your current location.

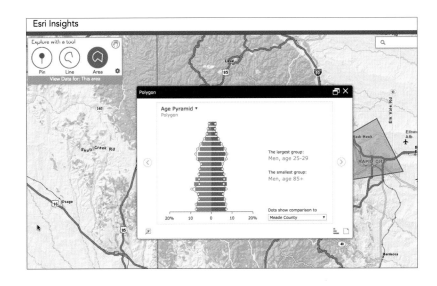

Integrate with your business data

Esri Maps for Office helps you integrate information from Microsoft Office products like Excel and PowerPoint with your web maps. See your Excel spreadsheet data mapped within the Excel environment and updated automatically as you work on your spreadsheet.

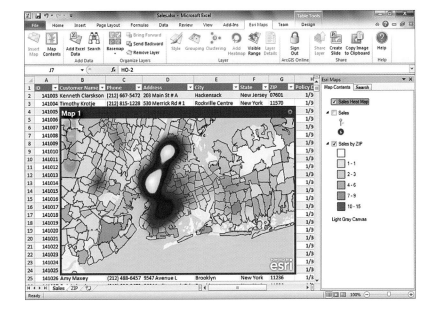

Quickstart

Use out-of-the-box apps, build apps without having to write any code, or code your own apps from scratch

▸ **Use Esri's ArcGIS apps**
These are out-of-the-box apps that are ready for you to use for your purposes:

Explore maps—Explorer for ArcGIS, take a tour
Collect data—Collector for ArcGIS, try collecting a damage assessment.
Manage operations—Operations Dashboard for ArcGIS, monitor a city's emergency response to an earthquake.
Analyze demographic data—Community Analyst, see an overview video of the app.
Analyze and evaluate scenarios—GeoPlanner for ArcGIS, get more information.
Integrate with your business data—Location Analytics, find the perfect app for your business.

▸ **Find apps in the ArcGIS Marketplace**
Esri has created a marketplace for you to find apps developed by Esri, business partners, and others, all built on top of ArcGIS.

▸ **Developers.arcgis.com**
If you know the difference between API and SDK, then head to the ArcGIS for Developers website or to GitHub.

▸ **Build your own app**
If the out-of-the-box apps don't do what you need, then why not build one yourself?

No coding required—Web AppBuilder for ArcGIS, make your first app in five minutes.
Configure template apps—ArcGIS Solutions, explore solution templates to jumpstart your projects.
Code a web app—ArcGIS API for Javascript, use the API to build your first web app.
Code a native app—ArcGIS Runtime SDKs, see the power of native apps with ArcGIS Runtime developer kits.

Share your apps

Web Apps
1. Select the app.
2. Configure it.
3. Save it and share it in ArcGIS Online or Server.

Native Apps
1. Find the app you want to share in the iTunes App Store or Google Play.
2. Share the URL with your audience.

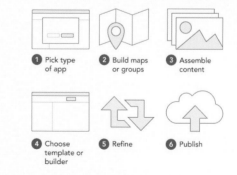

Learn ArcGIS Lesson

Use Web AppBuilder to create an Oso mudslide swipe map app

On March 22, 2014, a major landslide occurred in a semi rural area four miles east of Oso, Washington. A portion of an unstable hill collapsed, sending mud and debris across the North Fork of the Stillaguamish River, engulfing a neighborhood and and killing 43 people.

▸ Overview

As part of the Washington State team tasked with helping people understand what actually happened, your job is to create a before-and-after swipe map of the affected landscape.

In this lesson, you'll add before-and-after imagery layers to an ArcGIS web map. From there, you'll save the map, and then configure it in the

Web AppBuilder to craft a focused web app that utilizes several interesting tools like swipe and measurement.

▸ Build skills in these areas:
- Adding layers to a map
- Using Web AppBuilder
- Changing map symbols
- Sharing the map as a web application

▸ What you need:
- An ArcGIS Organizational Account
- Estimated time: 45 minutes

Start Lesson

Esri.com/ArcGISBookChapter7_Lesson

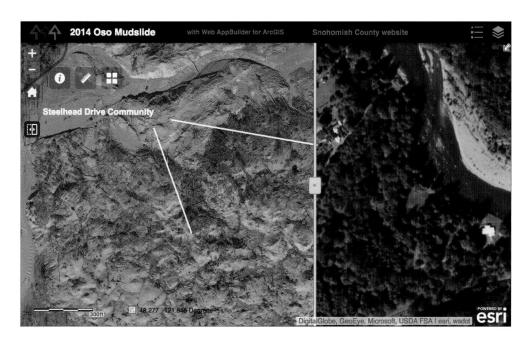

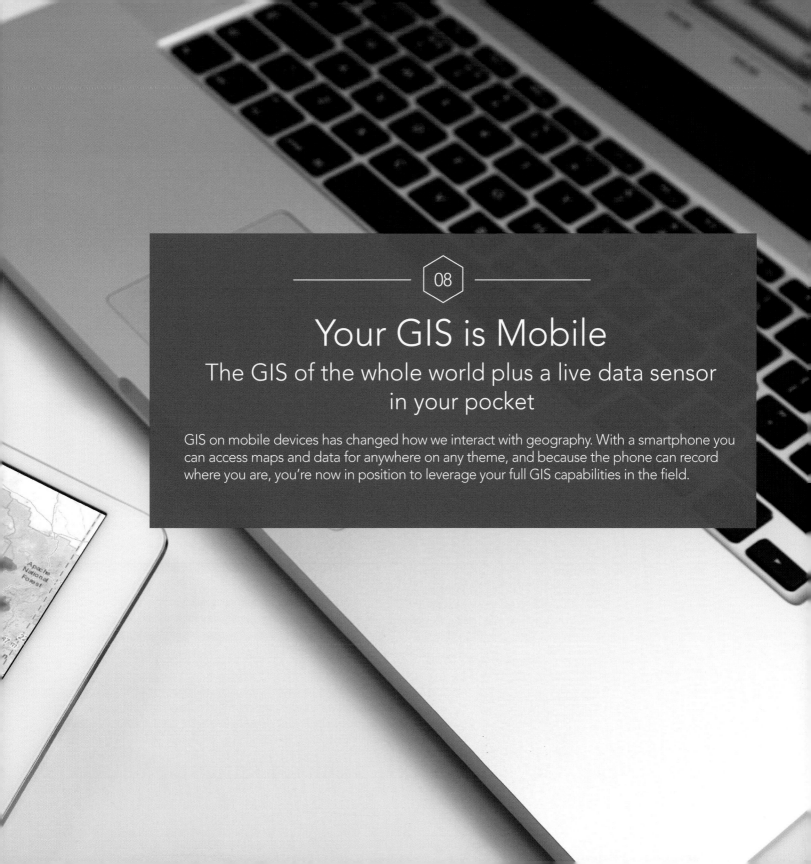

Your GIS is Mobile
The GIS of the whole world plus a live data sensor in your pocket

GIS on mobile devices has changed how we interact with geography. With a smartphone you can access maps and data for anywhere on any theme, and because the phone can record where you are, you're now in position to leverage your full GIS capabilities in the field.

GIS goes where you go

With mobile GIS, your GIS maps and apps go with you wherever you go. That's a big idea. The integration of the smartphone and GIS carries many implications in addition to the ones described here.

You can use your phone to capture geotagged photos and videos, and then use them to tell and share your stories. You can collect data in the field and update your enterprise information. Your phone can also be used to access enterprise information for your current location so that you have deeper knowledge and awareness.

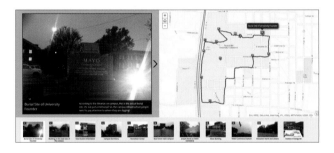

You can track and coordinate with members of your team in the field. You and the team can be guided to the right locations, and away from dangerous or otherwise undesired ones. You can receive triggers and signals when you enter certain zones or are approaching others.

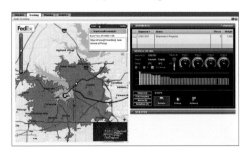

Your phone can access deeper geographic knowledge about any location.

Cities can notify their citizens about road and bridge closures. In turn, citizens can report on problems and issues and provide geotagged comments and feedback about proposals and plans.

You can be guided to exactly where you need to go—and not just along public roads but also private roads—to precisely where equipment is located, for example, and even inside of buildings and across campuses.

Clearly, your users can access, share, and apply a world of information about their location and activities. These come to life on your phone through a series of apps provided by the ArcGIS developer community as well as by Esri and its partners.

Thought Leader: Jeff Shaner

On the scene at the Deepwater Horizon oil spill

During the 2010 Deepwater Horizon oil spill in the Gulf of Mexico, I was part of a team sent by Esri to assist our customers among the several different emergency response agencies that were operating at the scene. The situation was somewhat intense and we were in meetings where a lot of information was flying around—not all of it accurate or timely. Dozens of teams were in the field—to monitor the developing situation, to collect data, and to conduct environmental surveys. The data collection effort was still largely paper-based and coordination among all of the teams was difficult.

The problem wasn't a lack of maps or GIS. These agencies were already among our most sophisticated users. The problem was in appropriately sharing and keeping each other updated as new information flowed into the operations center. I witnessed how the teams gathering data out on the water and along the shoreline were struggling early on to collect their information and to make it accessible so that it could be acted upon.

Within a week, thanks to a lot of hard work by scores of technologists, emergency response people, and staff from British Petroleum, many of the pieces were falling into place and we saw GIS beginning to be used for mobile data collection and communication. These teams began to share maps, data, videos, and photos, enabling responders to better coordinate with emergency command centers and ensure high levels of

Jeff Shaner is a Product Manager at Esri focused on the development of new mobile and web technology offerings.

situational awareness. As tragic as the event was, I did see how much more effective these teams were by using mobile GIS.

When we returned to Esri, we used this firsthand experience to help guide our product development forward, by applying many of the ideas that came out of those frantic weeks. One idea borne from this initiative became the first generation of our Collector app.

It's heartening for me to know that these same agencies are now equipped with Collector, among a whole suite of new applications that equip response teams with more efficient rescue and recovery capabilities.

What is Collector for ArcGIS?

The power of Collector enables organizations to use maps to gather data in the field and to synchronize the results with their enterprise GIS data. With Collector you can update data in the field, log your location, and put the data you capture back into your central GIS database directly from your phone or mobile device. This increases accuracy and helps eliminate recording errors. Fieldworkers are much more efficient and accurate, reducing error and time. And Collector increases the speed at which the information you collect in the field can be put to work throughout your organization.

You can download maps to your device to work offline; use GPS to create and update map data, points, lines, and area features; fill out easy-to-use map-driven forms; find places and get directions; track and report areas you visited—all these are functions of Collector.

Some ways that Collector is used
Anywhere that you see people doing work in the field there's a potential for the application of Collector.

Tree inventory

With mobile devices, tree inspectors working in suburban neighborhoods gather information about tree health and maintenance work.

Damage assessment

After the storm passes, damage assessment teams in tornado-prone Texas are deployed with tablets and smartphones to record its effects.

Water quality monitoring sites

Health and environmental departments inspect public waterways across the state of Florida to determine water quality status.

Survey123

Intelligent field polling for ArcGIS

It turns out that surveys are among the best ways to gather information from the field, and intelligent surveys are important to that process. Often the answers to one survey question determine additional follow-up questions. This ability to be adaptive is critical for sophisticated information gathering.

Survey123 has been used for intelligent information gathering on the spread of Ebola in West Africa and for recovery operations after the series of devastating Nepal earthquakes in 2015.

Survey123 provides an end-to-end workflow for designing and implementing surveys in the field, conducting them, and ensuring the results are geotagged and synchronized with your enterprise GIS.

Author the questionnaire

Conduct surveys

The landing site is at survey survey123.esri.com.

Case study: Gathering data in remote areas

Collecting field data in Canada's northern wilderness, with its abundance of marshes, rivers, rugged terrain, and dense forests, is always a challenge. Doing it using tablets—instead of the traditional paper-and-pen method—is a different kind of challenge altogether, involving transmitting data to the office from a remote region where Internet connection is sporadic.

As part of an environmental impact study on the effects of the construction of a proposed 430-km electric transmission line, three field crews from the study's lead consultant, Dillon Consulting, (Toronto) were sent out to gather information about environmental conditions along the proposed route. Using ArcGIS on mobile hardware allowed the system designers to put aside an antiquated, paper-based data collection process.

Instead of reporting the data when teams return from the field, analysts in the office are able to work with the data as it is being submitted. They can also send information back out to the crews, for example, as new sampling modifications are needed while conducting fieldwork.

Field crews gather data on tablets in three ways: by tagging maps, by entering information in text fields, and by taking photos or videos. The app seamlessly synchronizes this information with their enterprise GIS when Internet connections can be established after teams return from the field.

Standing in the rain in Northern Ontario, this fieldworker gathering data on environmental conditions is actually connected to an enterprise database.

Sometimes syncing occurs like this: a fieldworker can talk to an analyst on the phone, request a change—such as the color of a feature on a map—synchronize, and instantly see the update on her or his tablet.

Real-time syncing has other benefits, too. Data can be safely stored in the cloud, not just on field crew devices. If a tablet gets damaged during the project, no data will be lost.

The mobile GIS data flow

With a smartphone, everyone is a potential data sensor. In addition to its workforce applications, this opens a whole area called "VGI"(volunteered geographic information). Think of this as geolocated social media like tweets and blog posts. Even something as simple as a tweet can have location associated with it, and with geolocation, this content becomes part of a larger crowdsourcing effort.

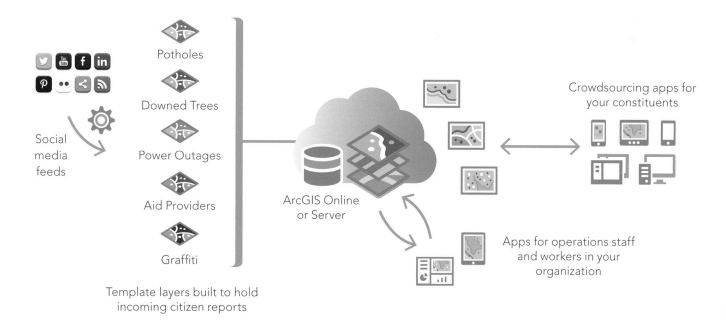

Social media feeds

Potholes

Downed Trees

Power Outages

Aid Providers

Graffiti

Template layers built to hold incoming citizen reports

ArcGIS Online or Server

Crowdsourcing apps for your constituents

Apps for operations staff and workers in your organization

Quickstart

Get on board with mobile GIS and Collector

▸ **Sign into your organization**
Log into your organization or the trial you started in the previous chapters.

▸ **Create a new map for collection**
Make a map for data collection by adding an editable layer to a new map. If you don't have Operations Dashboard for ArcGIS installed, install it from ArcGIS Online.

Note: The app downloaded from ArcGIS Online can be used to connect to your ArcGIS organization on either ArcGIS Online or Portal for ArcGIS. However, your company might require that you, instead, install from your Portal for ArcGIS. To

do so, search for Operations Dashboard on your portal, or contact your ArcGIS administrator.

▸ **Configure the data collection form**
Much like a paper form, the pop-up or digital form provides a structured way for users to enter information.

▸ **Share your map**
Share your map to a group to empower others to collect information.

▸ **Open the map on your phone**
Now that you're done, sign into Collector and view your map on your smartphone.

Create and share a map
Extend the reach of ArcGIS into the field and improve the accuracy and currency of your spatial data. Create and share maps for data collection in a couple of steps.

Go offline
Fieldwork takes you places without a data connection. Take your maps and data offline, collect data, and get back in sync once reconnected.

Collect data
Create the data that matters to your organization, from damage reports and service requests to places of historical interest. Easily include images and videos and share your work.

Jump-start with templates
Your industry has its own common datasets. Download templates to create the basic data structure you require and customize it for your particular job.

Learn ArcGIS Lesson

Manage a mobile workforce

You're the GIS specialist for the public works department of Naperville, a midsize Chicago suburb. Part of your job involves managing your field staff's routine inspections of fire hydrants. Inspections are currently done with pen and paper, so it takes a long time to transfer data to your GIS, and human error is frequent.

▸ **Overview**
In this project, you'll automate the process. You'll publish fire hydrant inspection data from ArcMap to ArcGIS Online. Then, you'll create a web map from the published layer, which you can share with your workforce. Lastly, you will use Collector for ArcGIS to input field observations directly into your web map, automatically updating it with the latest data.

▸ **Build skills in these areas:**
- Publishing a web layer
- Creating a web map
- Sharing maps with workers in the field
- Using Collector for ArcGIS

▸ **What you need:**
- ArcMap (Standard or Advanced license)
- Publisher or administrator role in an ArcGIS organization
- Collector for ArcGIS
- A smartphone or tablet with iOS 7 (or a later version) or Android 4.0 (or a later version)

Start Lesson

Esri.com/ArcGISBook/Chapter8_Lesson

Real-Time Dashboards
Integrating live data feeds for managing operations

Real-time dashboards provide a way to absorb and make meaning from the torrent of real-time information that is used to drive so many decisions. Dashboards are your secret weapon for visualizing and putting meaning behind all of these real-time feeds.

How real-time dashboards are used

A vast amount of data is created every day from sensors and devices: GPS devices on vehicles, objects, and people; sensors monitoring the environment; live video feeds; speed sensors in roadways; social media feeds. What it means is that we have an emerging source of valuable data. It's called "real-time" data. Only recently has the technology emerged to enable this real-time data to be incorporated into GIS applications.

The real-time GIS capabilities of the ArcGIS platform have transformed how information is utilized during any given situation. Real-time dashboards provide actionable views into the daily operations of organizations, empowering decision makers and stakeholders with the latest information they need to drive current and future ideas and strategies. Dashboards answer questions like: What's happening right now? Where's it happening? Who is affected? What assets are available? Where are my people?

Some applications of real-time dashboards

- Local governments use real-time information to manage operations such as tracking and monitoring snowplows and trash trucks.

- Utilities monitor public services including water, wastewater, and electricity for consumers.

- Transportation departments track buses and trains and monitor traffic flows, road conditions, and incidents.

- Airport authorities and aviation agencies track and monitor air traffic worldwide.

- Oil and gas companies track and monitor equipment in the field, tanker cars, and field crews.

- Law enforcement agencies monitor crime as it happens, as well as incoming 911 calls.

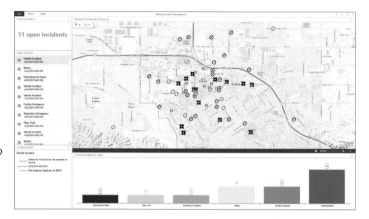

The Redlands Incident Management operations view provides a way to monitor vehicle locations, shelter capacity, and incidents in real time using dynamically changing data sources and information displays.

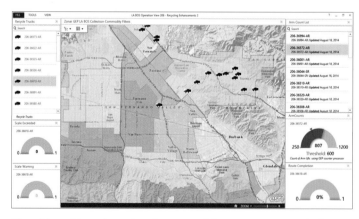

The City of Los Angeles Bureau of Sanitation uses dashboards to track and monitor sanitation trucks throughout the city.

- So they can issue early warnings and reports, federal agencies like the Federal Emergency Management Agency (FEMA), the US Geological Survey (USGS), the National Oceanographic and Atmospheric Administration (NOAA), and the Environmental Protection Agency (EPA) gather vast amounts of information about the environment. They monitor weather, air and water quality, floods, earthquakes, and wildfires.

- Companies use real-time social media feeds like Twitter to gauge feedback and monitor social sentiment about particular items of interest.

- Individuals use smartphones, smartwatches, smart sensors, radio-frequency identifications (RFIDs), beacons, fitness bands, etc., to capture and visualize information about everything we do (often referred to as the Internet of Things).

- Emergency management agencies monitor public safety during large events, such as marathons and the Olympics.

A real-time dashboard tracks athletes and incidents throughout the IRONMAN World Championship in Hawaii.

How *real* is real time?

Real-time data is as current as the data source that is updating it, whether that data is being updated every second, minute, hour, or daily. What is real time to one organization might not be real time to another, depending on the type of scenario being monitored.

Real time is a concept that typically refers to the awareness of events at the same rate or at the same time as they unfold (without significant delay). Very often it's confused with frequency, or interval between events, which is essentially how often the event is updated. The update interval, or frequency, relates to the term "temporal resolution," which can vary from one application to another.

For example, most aircraft monitoring systems provide two updates for every second, while it may take every hour to provide a weather update. For monitoring their networks, energy utilities use systems, also known as SCADA (Supervisory Control and Data Acquisition), that sample data about voltage, flow, pressure, and more from analog devices at very high frequencies (e.g., 50 hertz). This can result in high resource requirements for such things as network bandwidth, system memory, and storage volume.

The data that fueled geographic applications in the past was created to represent the state of something at a specific point in time; data captured for what has happened, or what is happening, or what will happen. While this GIS data is valuable for countless GIS applications and analyses, today the current snapshot of what is happening now falls out of sync very quickly with the real world, becoming outdated almost as soon as it is created.

What is real-time GIS?

Real-time GIS can be characterized as a continuous stream of events flowing from sensors or data feeds. Each event represents the latest state, including position, temperature, concentration, pressure, voltage, water level, altitude, speed, distance, and directional information flowing from a sensor.

As you'll discover in this chapter, maps provide one of the most basic frameworks for viewing, monitoring, and responding to real-time data feeds.

Aspects of a real-time GIS

Acquire real-time data

A utility organization may want to visually represent the live status of its network with information that is captured by sensors in the field. While the sensors on the network are not physically moving, their status and the information they send changes very rapidly. Radio-frequency identification (RFID) is being used in a wide variety of environments to keep track of items of interest. Warehouses and logistics companies use RFID to track and monitor inventory levels. Hospitals use it to track equipment to make sure it has gone through proper cleaning procedures before entering another patient's room.

A wide range of real-time data is accessible today. Connectors exist for many common devices and sensors enabling easy integration into your GIS.

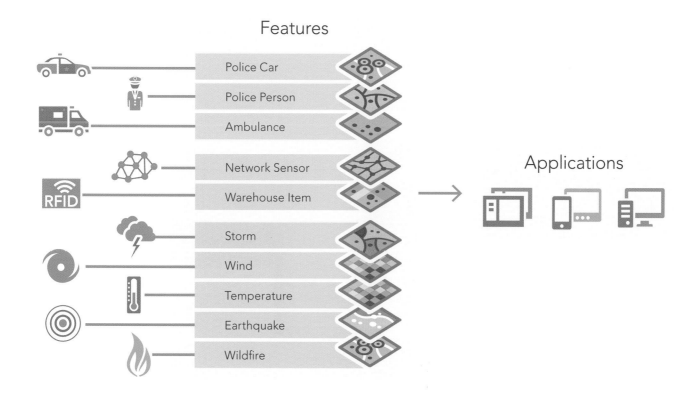

Features

Police Car
Police Person
Ambulance
Network Sensor
Warehouse Item
Storm
Wind
Temperature
Earthquake
Wildfire

RFID

Applications

Perform continuous processing and analysis

While it's easy to connect a sensor to a feature and display it in real time, the next thing you might want to do is perform some additional processing or analysis on that data as you receive it. For instance, assume you are a member of an organization that is responsible for tracking and monitoring fishing vessels to ensure they are not fishing in a restricted zone. To accomplish this, you need an Automatic Identification System (AIS) feed that is updating a vessel feature with current locations. At the same time, you must continuously process, or filter, a vessel's current location to detect when it enters a restricted zone, so that when it does, an alert appears on the operator's dashboard. Having the ability to perform continuous processing and analysis on data as it is received allows you to detect such patterns of interest in real-time.

Communicate the results

Another example of continuous analysis on a real-time stream of data is one in which parents want to be notified when their child has left school property. Continuously updating the child's current location, the analysis detects when he or she leaves school property. At that moment, the parents are immediately notified by email and SMS text message.

Sending these kinds of updates and alerts to those who need it, where they need it, allows your key stakeholders to be notified in a way that is convenient for them.

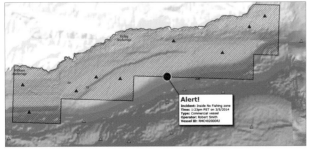

Components of a real-time dashboard

Real-time dashboards are created by adding "widgets" to an operation view. Operation views are easy to set up and configure. The map widget creates the primary map display and serves as the source of data for other widgets. You choose which data source or attribute value is displayed by the widget, specify the appearance settings, enter a description or explanatory text, and set any other properties required for the particular widget.

Widgets are used to represent your real-time data in a visual way. For example, a symbol could represent the location of a feature on a map; a text description could be displayed in a list; and a numerical value could be shown as a bar chart, a gauge, or an indicator.

Each operational view is updated with the latest data by setting a refresh interval on both the widget and each layer.

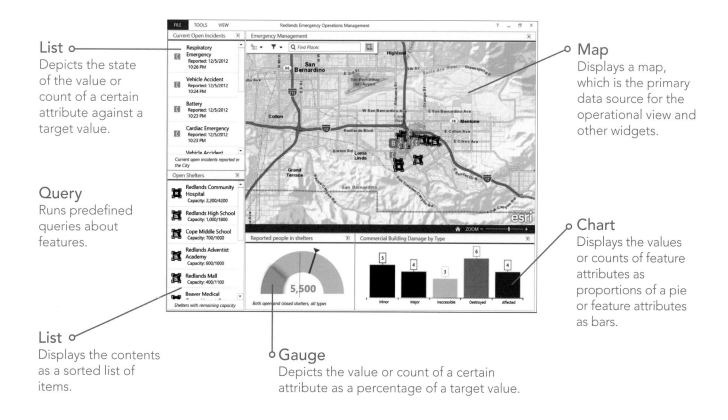

List
Depicts the state of the value or count of a certain attribute against a target value.

Query
Runs predefined queries about features.

List
Displays the contents as a sorted list of items.

Gauge
Depicts the value or count of a certain attribute as a percentage of a target value.

Map
Displays a map, which is the primary data source for the operational view and other widgets.

Chart
Displays the values or counts of feature attributes as proportions of a pie or feature attributes as bars.

Real-time GIS platform capabilities
Working with real-time data

ArcGIS GeoEvent Extension for Server is an extension to ArcGIS for Server that brings your real-time data to life, allowing you to connect to virtually any type of streaming data, process and analyze that data, and send updates and alerts when specified conditions occur, all in real time.

Your everyday GIS applications become frontline decision applications, helping you respond faster with increased awareness whenever and wherever change occurs.

Connecting to feeds
The GeoEvent Extension is capable of receiving and interpreting real-time data from virtually any source. The system understands how the real-time data is being received as well as how the data is formatted. Input Connectors (shown here) allow you to acquire real-time data from a variety of sources.

Sending updates and alerts
Output Connectors are responsible for preparing and sending the processed data to a consumer in an expected format. An Output Connector translates its events into a format capable of being sent over a particular communication channel.

Performing real-time analytics
GeoEvent services enable you to define the flow of the event data as well as to add any filtering and processing on the data as it flows to the Output Connector. Applying real-time analytics allows you to discover and focus on the most interesting events, locations, and thresholds for your operations.

Send updates and alerts

Acquire real-time data

Operations Dashboard for ArcGIS

GeoEvent Extension

ArcGIS Server

Process and analyze real-time data

Visualizing real-time data

With the Operations Dashboard you can create real-time dashboards that allow you to visualize and display key information about your operations. These operational views can be stored in your ArcGIS organizational account and shared with members of your organization, with groups within your organization, as well as publicly with anyone who has an ArcGIS organizational account.

Real-time data storage

In most cases, data streamed into ArcGIS in real-time will be captured in a geodatabase. To support historical archiving of events, a best practice is to use another feature class (also known as a historical or temporal feature class) to store all of the events received from the data. This allows the state of each object to be stored indefinitely, everything from the first event received until now. As you can imagine, the size of this data can grow to be large, especially over an extended period of time. The growth rate of your data is largely dependent on the message size and the frequency of the incoming data. A best practice is to define and enforce a retention policy for how much history is maintained active in the geodatabase.

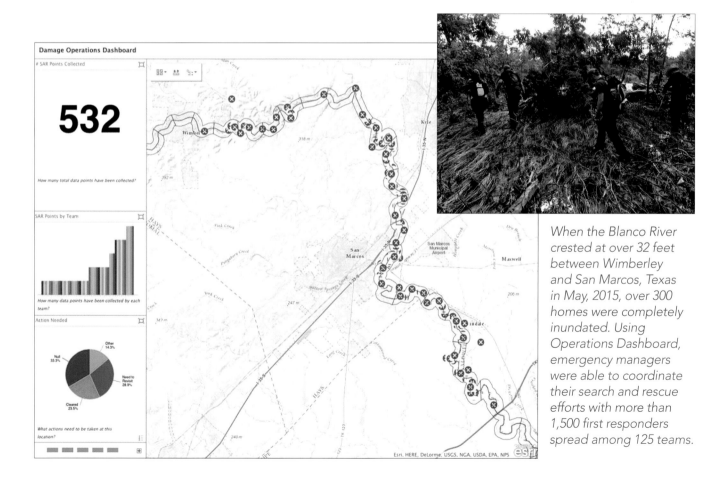

When the Blanco River crested at over 32 feet between Wimberley and San Marcos, Texas in May, 2015, over 300 homes were completely inundated. Using Operations Dashboard, emergency managers were able to coordinate their search and rescue efforts with more than 1,500 first responders spread among 125 teams.

Examples of real-time data sources

Real-time data takes on many different forms and has many different applications. Some of these examples link to live feed maps, some to the item information pages for the feed itself, and some to the developer API.

Active hurricanes

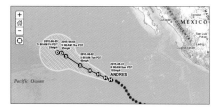

This map service describes the current and forecasted paths of tropical activity from the National Hurricane Center.

Hourly wind conditions

The Current Wind Conditions layer is created from hourly data provided from NOAA.

USGS earthquakes

Minute-by-minute earthquake data for the last 90 days comes from the USGS and contributing networks.

L.A. Metro bus locations

L.A. Metro's Realtime API gives access to the positions of Metro vehicles on their routes in real-time.

Stream gauges

These stream gauge feeds allow users to map current water levels to monitor flood and drought risk.

World traffic

Updated every five minutes, this dynamic map service monitors traffic speeds and incidents.

Twitter feeds

ArcGIS provides a sample for displaying geolocated tweets on a web map.

Instagram feeds

ArcGIS provides a sample for displaying geolocated Instagram posts on a live web map.

Severe weather

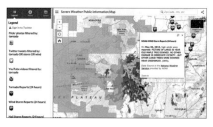

This map features live feed layers for severe weather across the United States and Canada.

Case study: The 119th Boston Marathon

After the horrific bombing that took place at the 2013 Boston Marathon, the safety of spectators and participants at the 2015 event was of foremost concern. With staff from the Massachusetts Emergency Management Agency, Massachusetts State Police, and others, the operations center was equipped with wall-size monitors to display key information about various aspects of the race, including latest status, weather, news, and street camera video feeds.

As the race started, it was evident that the real-time dashboard was of particular interest. It captured the pulse of the event by integrating key information, such as the density of the 30,000 participating runners at each segment of the course (Start, 5k, 5k–10k, etc.), progress along the course, and the percentage complete. Additionally, the dashboard tracked the location of each emergency vehicle supporting the race, medical-related incidents in 26 medical tents, emergency calls and their status, and live weather information. All of this was synthesized into a real-time dashboard. Everyone in the room (and those in satellite operations centers) was able to visualize what was happening, as it occurred.

This effective flow of information helped contribute to a successful event, which was, of course, the goal.

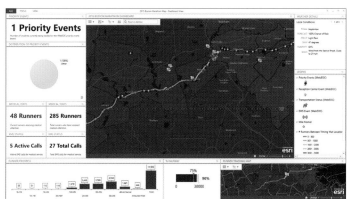

The Boston Marathon real-time dashboard consists of a map of the course, the number of runners in each section of the course, weather, emergency calls, and status of injured participants.

Quickstart

Get your real-time dashboard up and running

Operations Dashboard for ArcGIS is a Windows application that you can download and run locally. It's where you design your operational views.

1. Download and install Operations Dashboard for ArcGIS

2. Documentation is available online.

3. An ArcGIS organizational account is required; if you do not have one see page 15.

Tips for real-time dashboards

There are a number of principles to consider when configuring real-time dashboards:

- Design it for a specific purpose or scenario.

- Keep it easy to understand and intuitive, so no one needs to ask for explanations.

- Make the layout simple so that it focuses attention on the most important information.

- Present the information in a prioritized way that assists in making timely decisions.

- Render it flexible enough to be delved into for more detail when necessary.

- Make sure it provides timely updates and synchronizes all the widgets in real time.

Multi-display vs single display dashboards

Operations Dashboard for ArcGIS provides two kinds of operations views:

- Multi-display operations views are useful in environments with multiple monitors (like in a desktop setting). They are especially useful when you have a centralized operations center where staff are collectively viewing multiple monitors displaying continually updated maps, charts, and video feeds.

- Single-display operations views are designed for individuals to use on mobile phones, tablets, and web browsers.

GeoEvent Extension for Server

This extends ArcGIS for Server and provides capabilities for consuming real-time data feeds from a variety of sources, continuously processes and analyzes that data in real-time, and updates and alerts your stakeholders when specified conditions occur.

To learn more about the GeoEvent Extension, access the documentation, sample connectors, and videos at links.esri.com/geoevent.

Learn ArcGIS Lesson

Explore then recreate a dashboard

In an emergency, a coordinated effort saves time, which can save lives.

▸ Overview

For this lesson, assume you are a GIS coordinator at an emergency response agency that is responsible for managing local emergency vehicles for the City of Redlands, California. You have at your disposal a lot of information and data sources to track. You need a way to bring all of this information together into a dashboard that will enable police and fire supervisors to know where all their assets are in relation to actual incidents happening on the streets.

In this lesson, you explore an existing real-time dashboard that has been configured around this response effort. Then you will create an instance of the dashboard. (The scenario and the data you work with in this lesson are simulated and not based on actual events.)

▸ Build skills in these areas:

- Create a web map
- Create a real-time dashboard
- Work with real-time data
- Customize an operations view

▸ What you need:

- ArcGIS organizational account
- Operations Dashboard (for Windows only)

Start Lesson

Esri.com/ArcGISBook/Chapter9_Lesson

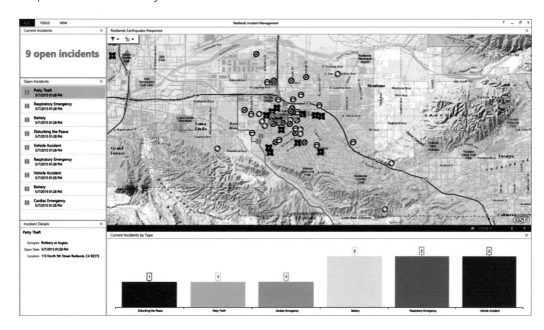

10

GIS is Social
Web GIS is the GIS of the world

Your own GIS is simply your view into the larger system. It's a two-way street. You consume information that you need from others, and in turn, you feed your information back into the larger ecosystem.

GIS is collaborative
Geography is key for integrating work across communities

Modern GIS is about participation, sharing, and collaboration. As a Web GIS user, you require helpful, ready-to-use information that can be put to work quickly and easily. The GIS user community fulfills that need—that's the big idea. GIS was actually about open data long before the term gained fashion because the people who were doing it were always looking for ways to deepen and broaden their own GIS data holdings. No one agency, team, or individual user could possibly hope to compile all the themes and geographic extents of data required, so people networked about this to get what they needed.

Since the early days in GIS, people realized that to be successful they would need data from other sources beyond their immediate workgroups. People quickly recognized the need for data sharing. Open GIS and data sharing gained traction quite rapidly across the GIS community, and continue to be a critical aspect in GIS implementation.

With cloud computing and the mobile/app revolution, the GIS community is expanding to include almost anyone on the planet. The data in every GIS is being brought together virtually to create a comprehensive GIS of the world, and nearly everyone can take GIS with them everywhere they go on their tablets and smartphones. Geography and maps enable all kinds of conversations and working relationships both inside and outside your organization.

GIS is for organizations
First and foremost, your GIS can be used by people throughout your organization. In Web GIS, maps are purpose-driven and their intended audience may include your executives, managers, decision makers, operations staff, field crews, and constituents. ArcGIS Online enables you to extend your reach to these users.

GIS is for communities
GIS users collaborate across communities. These communities may be based on relationships fostered by living in the same geography (a city, region, state, or country) or by working in the same industry or subject matter (conservation, utilities, government, land management, agriculture, epidemiology, business, etc.). In these communities, users share critical data layers as well as map designs, best practices, and GIS methods.

GIS is for public engagement
People everywhere are starting to engage with GIS. They have been using maps as consumers, and now they are interested in applying them at work and in their community relationships. Often this involves communicating with the public by telling stories using maps. More and more, members of the public are providing input and collecting their own data for GIS organizations and the public good. This makes for better civic engagement at multiple levels.

GIS work is a valued profession
Community is vital in GIS

It is vital to recognize the phenomenal growth of GIS in people's lives and how its effect extends beyond its economic and fiscal impacts. You are— or can be—an active participant in a truly amazing field. Every day, millions of people are using GIS in government, industry, and academia. Even smaller organizations are hiring dedicated GIS professionals to improve the quality and accuracy of work being accomplished, and the benefits are measurable. GIS helps people make better decisions, reduce costs, work more efficiently, communicate better, and gain key insights.

Gathering of the faithful at the 2014 International Esri User Conference. The UC has been an annual event since 1981.

Globally, GIS and the related geospatial economy is valued at more than $250 billion per year. The geospatial segment is one of the fastest growing in the tech field overall. That's saying something because everyone knows how fast tech is growing. This segment is considered by the US Department of Labor to be one of the three technology areas that will create the greatest number of new jobs over the next decade. It's growing at 35% overall, and some of the sectors like business GIS are growing at a 100% clip.

This worldwide community is busy every workday implementing GIS: They are growing their expertise and extending their reach throughout organizations and communities all over the planet. The work these people do is impossible to pigeonhole because it is so broad, yet it does tend to focus its attention on critical resource issues, environmental collapse, climate change, and other daunting problems. The most hardcore users of GIS tend to be passionate and interested in the world, and dedicated to making a difference. To them, and maybe to you, its important to feel like the work you do means something.

GIS is also face-to-face
For a lot of the reasons described in this book, GIS is a collegial profession with a strong networking aspect. Organizations like URISA, American Association of Geographers (AAG), and others have long held well-attended conferences. From the beginning, Esri has encouraged face-to-face networking of its user community through regional user groups, industry user groups, its Developer Summit, and its annual Esri International User Conference. With more than 16,000 in attendance, this event is the largest GIS gathering annually. To sit in the audience during the plenary session is to feel part of something truly larger than yourself.

ArcGIS is for organizations

GIS has a critical role to play in your organization. ArcGIS is a geography platform enabling you to create, organize, and share geographic information in the form of maps and apps with workers throughout your organization. These run virtually anywhere—on your local network or hosted in the ArcGIS Online cloud. The maps and apps that you share are accessible from desktops, web browsers, smartphones, and tablets.

The role of the GIS department

Professional GIS provides the foundation for GIS use across your organization. It all starts with the work you do on your professional GIS desktops. You compile and manage geographic data, work with advanced maps, perform spatial analysis, and conduct GIS projects. Your resulting GIS content can be put to use by others in countless ways. Your work is shared as online maps and apps that bring GIS to life for users within your organization and beyond.

Portals enable collaboration across your organization

A key component in your organizational GIS is your information catalog or Portal. This catalog contains all the items that are created, and ultimately shared, by your group's users.

Every item is referenced in your organization's information catalog—your Portal. Each contains an item description (often referred to as metadata) and any item can be shared with selected users within as well as outside your organization.

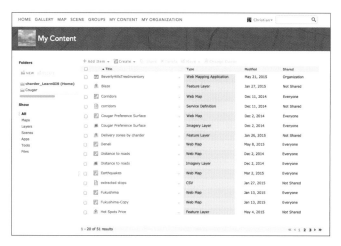

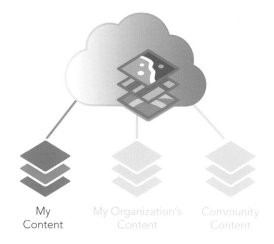

Your Portal contains the gallery of GIS items that are used by people in your organization. These include items such as maps, layers, analytical models, and apps.

ArcGIS provides intelligent online content management that enables you to create and share useful maps and apps with your users. You can engage ArcGIS to organize and distribute your geographic information and tools. With a portal, certain users will have access to apps that support their specific work tasks, like apps configured for collecting data in the field. Some maps will be shared with the entire organization, like the basemaps that provide the foundation for all the work performed across your organization. Some users will create their own maps by mashing up their data layers. And some items, like story maps about your organization's work, might be shared with everyone including the public.

Access to GIS content across your organization

1. Start with you and your organization's GIS content.

2. Combine it with community information layers shared by other users with whom you collaborate or shared by the broad GIS community.

3. Create maps and analytical tools for your users and constituents, and share these online.

4. Share your maps and geographic information layers with others throughout your organization and, optionally, beyond.

GIS roles

GIS is about the people in your organization and the purpose-driven maps and apps they apply to do their work. Every user is given an ArcGIS account (i.e., a login) and assigned a role for using ArcGIS. For example:

- Administrators manage the system and enable new users to participate by granting privileges for their roles in your organization's GIS. There are only a handful of administrators (one or two) in each organization.

- Publishers create maps and app configurations that can then be shared with users throughout your organization and with the public. Publishers also help to organize content by creating and managing logical collections or groups. Users throughout your organization find their maps and apps in these logical groups.

- Users create and use maps and apps and then share them with others—inside and outside your organization.

Geodesign
Using social engagement in community planning

"Everyone designs who devises courses of action aimed at changing existing situations into preferred ones."
-Herbert Simon, political scientist (1916-2001)

Geodesign provides a planning methodology and approach to project design and decision-making, and it is best practiced by a community of collaborators. A technical design approach is also involved. Once objectives for a project are articulated, professionals survey and characterize a landscape. They identify its special resources and the opportunities to support a project as well as the constraints that limit what might be possible or practical. GIS is often used in this phase to perform suitability/capability analysis. These results are used to generate the landscape of opportunities and constraints. Subsequently, design alternatives are sketched onto the landscape, and further GIS analysis is used to evaluate, compare, and analyze the various design alternatives.

The practice of geodesign requires collaboration among project participants. The most important aspect is the feedback and ideas that are generated by the participants—including local citizens and stakeholders who may be affected. Most geodesign activities are about this kind of community engagement and consideration. GIS provides a useful tool for others to participate in the evaluation by providing the ability to consider the issues of other stakeholders.

Many problems in the world are usually not well defined, not easily analyzed, and not easily solved. What we do know is that the issues are very important and require thoughtful consideration. They are beyond the scope and knowledge of any one person, discipline, or method. People must begin to understand the complexities, and then figure out ways to collaborate. Collaboration is a common thread, and social benefits are the central objective.

Geodesign, as an idea, has the potential to enable more effective collaboration between the geographically oriented sciences and the multiple design professions. It is clear that for serious societal and environmental issues, designing for change cannot be a solitary activity. Inevitably, it is a social endeavor.

—Adapted from _A Framework for Geodesign: Changing Geography by Design_ by Carl Steinitz

A Framework for Geodesign
Carl Steinitz

THE PEOPLE OF THE PLACE

GEOGRAPHIC SCIENCES

DESIGN PROFESSIONS

INFORMATION TECHNOLOGIES

Changing Geography by Design

Thought Leader: Clint Brown

GIS is participatory

It's no secret that modern GIS is participatory. All of us need access to other people's data to do our jobs and to do them well. And we've all worked for years on building communities that share information. Millions of users around the world have already built their data layers in hundreds of thousands of organizations. They continue to build them for their areas of interest, for their geographies, and for their key themes of information. The GIS community is building these layers at all levels of geography, from the community level to the regional level, at state and country levels, and even at global levels.

And what's interesting about geographic data is that it's all about layers. So there's this global collection of information that's been assembled and continues to be kept current and further built out by the GIS user community—all in a series of layers that reference onto the Earth, which makes it very easy to integrate that information.

Meanwhile, GIS is moving to the cloud, to this big network of computers that makes information available. One of the things that's happening with Web GIS is that every layer has a URL—it's got an address that's findable and usable. You can reference a data layer and begin to use it, apply it, and bring it into your GIS work.

So GIS provides a kind of integration engine, and this is a profound idea. By all of us focusing and working on our own geographic information systems, we're assembling this very promising, comprehensive GIS of the world. This is growing every day; it's getting richer every day. The cloud systems for using

Clint Brown is Director of Software Products at Esri. Early in his career as a GIS analyst in Alaska, he recognized the effectiveness of working together with analysts at other agencies.

Web GIS are becoming more capable, enabling all of us to have access to these rich collections of information. Now certainly, it's not like everybody uses all the millions of layers. We still work with what's relevant to us and what's appropriate for us to use. But when you're doing your work, you're beginning to realize you have access to other people's layers that are very important and very relevant to the work you're doing.

 Watch a video discussion with Clint Brown
Esri.com/ArcGISBook/Chapter10_Video1

Social GIS and crowdsourcing

Many GIS organizations are discovering compelling and useful ways to engage with their local communities. Geography and maps facilitate such civic engagement.

Citizen reporting

Citizens in your city can report on problems and situations they encounter daily.

Gathering feedback on proposed zoning changes and development plans

Polling apps enable you to share project information and to solicit feedback. This app enables citizens to submit their feedback about proposals of any scale or size in their community.

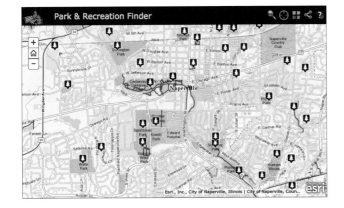

Finding government services

Local government entities of all sizes can provide maps and information to citizens about key services and events in their neighborhoods.

The rise of community engagement

Citizens and constituents everywhere are beginning to embrace GIS and engage with local GIS organizations. Here are a few examples of community engagement.

Humanitarian aid

Direct Relief, a global nonprofit providing medical assistance to people affected by poverty and emergency situations, works with Syrian refugees on a mission to treat and document skin disease among refugees. The NGO supplied each medical record scribe with a tablet equipped with the Survey 123 app. Despite the remote nature of the refugee camps and limited time, the scribes were able to easily use the app as part of the triage process.

The Dustcam Dashboard

Great Basin Unified Air Pollution Control District is a California agency that works to protect the people in its jurisdiction from the harmful effects of air pollution. With the ongoing California drought, the problem of respiratory disease caused by dust storms is on the increase. This story map collects all the webcam feeds onto a single map to view the current dust storm conditions.

White House Science Fair

A storytelling app, currently in prototype, will enable authors to easily configure crowdsourced and citizen science projects. Participants will be able to sign in using their social media identities and upload photographs, descriptions, and quantitative data. Administrators will be able to vet submissions. Projects can range from local, classroom-based projects to global, public-facing initiatives.

Quickstart

Set up your organizational account

▸ **Start the ArcGIS Trial**
From your web browser, visit the ArcGIS Trial page. Fill out your name and email address and click Start Trial.

▸ **Activate the ArcGIS Trial**
Open your email and follow the instructions from the Esri email to activate your ArcGIS Online account. This account, for which you will be the administrator, will allow you and four others to use ArcGIS Online.

▸ **Set up your organization**
There is one more step before your account is activated. Think carefully about your or your organization's short name because this will form the URL for you or your organization (and eventually all your content). Click Save and Continue.

▸ **Get inspired at the Apps page**
To get the Desktop software or other apps, go to the Free Trial page which is always accessible under your account name in ArcGIS.

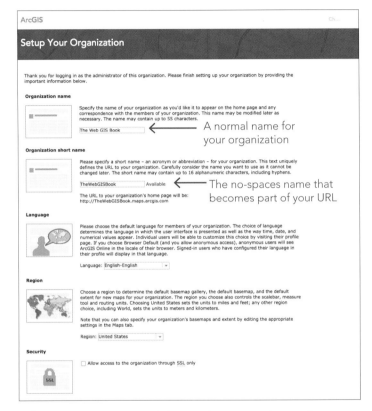

When you set up your trial, pay close attention to these two options. They form your organization's identity as others will see it online.

ConnectED

An education initiative

In response to President Barack Obama's call to help strengthen STEM education through the ConnectED Initiative, Esri President Jack Dangermond announced that Esri will provide a grant to make the ArcGIS system available for free to the more than 100,000 elementary, middle, and high schools in the United States, including public, private, and home schools.

ConnectED is a US government education program developed to prepare K-12 students for digital learning opportunities and future employment. The Initiative sets four goals to establish digital learning in all K-12 schools in the United States during the

next few years. These goals include high-speed connectivity to the Internet, access to affordable mobile devices to facilitate digital learning anytime, anywhere, high-quality software that provides multiple learning opportunities for students, and relevant teacher training to support this effort.

Watch a video by the inimitable Joseph Kerski
Esri.com/ArcGISBook/Chapter10_Video2

Go to ConnectED

www.esri.com/ConnectED

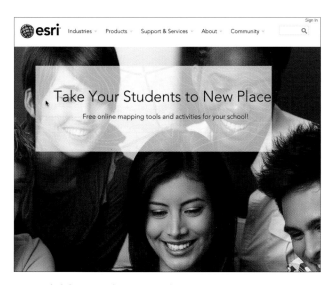

Your children and your students can engage in learning and applying ArcGIS using this book and the exercises it contains. This is a great way to get your students started with ConnectED.

What's your big idea?

Whatever it is, we thank you for becoming engaged in the ArcGIS experience. We are humbled by the work of our users around the world. Our work is about serving them—and hopefully, you as well.

About Esri

Esri is an exciting company doing important work. Our technology enables organizations to create responsible and sustainable solutions to problems at local and global scales.

At Esri, we believe that geography is at the heart of a more resilient and sustainable future. Governments, industry leaders, academics, and nongovernmental organizations (NGOs) trust us to connect them with the analytic knowledge they need to make these critical decisions that shape the planet.

We invite you to discover ways that you can leverage our technology and expertise in your own organization.

Esri vision video

Esri.com/ArcGISBook/Chapter10_Video

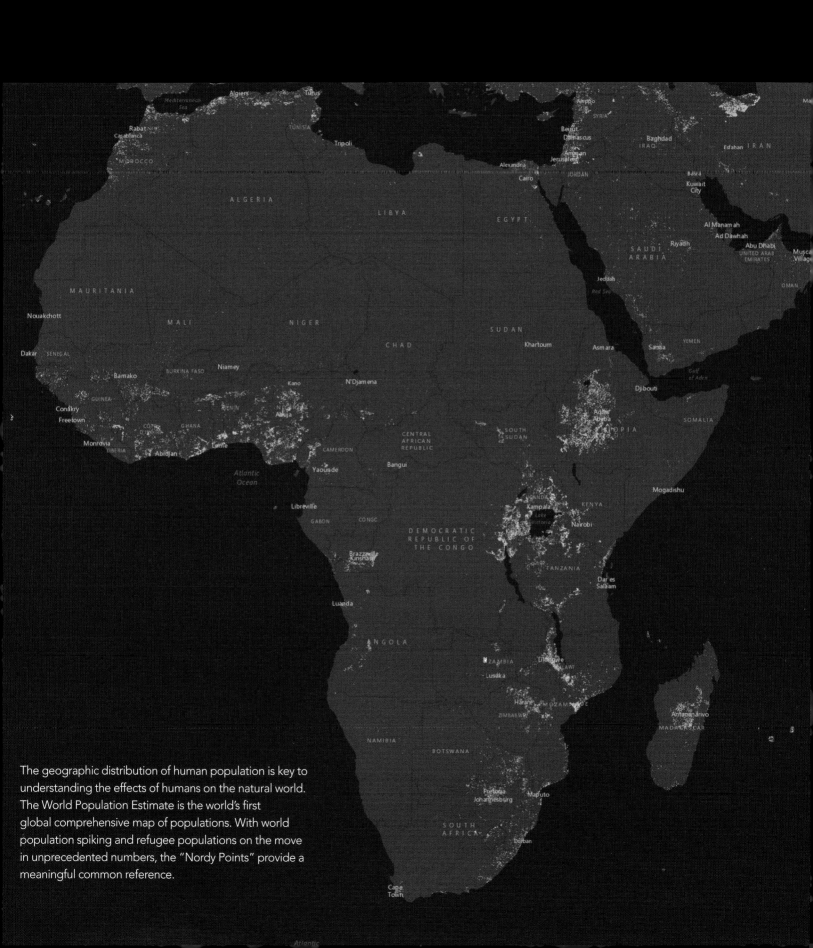

The geographic distribution of human population is key to understanding the effects of humans on the natural world. The World Population Estimate is the world's first global comprehensive map of populations. With world population spiking and refugee populations on the move in unprecedented numbers, the "Nordy Points" provide a meaningful common reference.

Resources for further study

Learn.ArcGIS.com

Esri Press

ArcGIS Solutions

GeoNet Community

Esri Training

ArcGIS Marketplace

Smart Communities

ArcGIS for Desktop

ArcGIS Support

Further reading, books by Esri Press

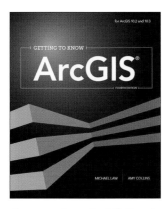

Getting to Know ArcGIS, Fourth Edition by Michael Law and Amy Collins

Making Spatial Decisions Using GIS: A Workbook, Second Edition by Kathryn Keranen and Robert Kolvoord

Making Spatial Decisions Using GIS and Remote Sensing: A Workbook, Second Edition by Kathryn Keranen and Robert Kolvoord

Getting to Know Web GIS by Pinde Fu

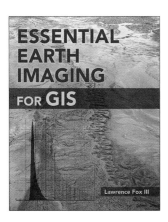

Essential Earth Imaging for GIS by Lawrence Fox III

The GIS 20: Essential Skills, Second Edition by Gina Clemmer

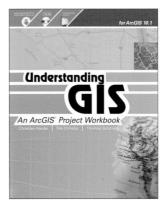

Understanding GIS: An ArcGIS Project Workbook, Second Edition by Christian Harder, Tim Ormsby, and Thomas Balstrøm

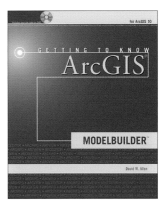

Getting to Know ArcGIS ModelBuilder by David W. Allen

Available on the Esri Books app; for more information visit esri.com/bookstoreapp.

See esripress.esri.com/bookresources for more books, exercises, data, software, and updates for Esri Press titles.

Contributors & acknowledgments

Contributors

Volume Editor: Christian Harder
Project Director: Clint Brown
Chapter 1—Christian Harder, Clint Brown
Chapter 2—Mark Harrower, Clint Brown
Chapter 3—Allen Carroll, Rupert Essinger
Chapter 4—Christian Harder, Tamara Yoder
Chapter 5—Linda Beale, Andy Mitchell
Chapter 6—Nathan Shephard
Chapter 7—Will Crick, Justin Colville
Chapter 8—Paul Barker, Jeff Shaner
Chapter 9—Greg Tieman, Morakot Pilouk
Chapter 10—Christian Harder, Clint Brown

Acknowledgments

Project managing:	Kathleen Morgan
	Riley Peake
Editing:	Candace Hogan
	Dave Boyles
Book design and layout:	Steve Pablo
Website design:	Karsten Thorson
Product planning:	Sandi Newman
Print production:	Lilia Arias

The work of a number of Esri cartographers is featured. Thanks to Kenneth Field, Andrew Skinner, Damien Saunder, Wesley Jones, Michael Dangermond, Jim Herries, Charlie Frye, Owen Evans, and Richie Carmichael.

The Learn ArcGIS team is Aileen Buckley, Bradley Wertman, Christian Harder, John Berry, Kevin Butler, Kyle Bauer, Nancy Morales, and Tim Ormsby.

Special thanks to Catherine Ortiz, Tammy Johnson, Kelley McKasy, Stefanie Tieman, Gisele Mounzer, Patty McGray, and Cliff Crabbe for their support throughout the project; to Brian Sims and Craig McCabe for their advice and input; and to Joseph Kerski, Deane Kensok, Sean Breyer, Adam Mollenkopf, Adam Buchholz, Jessica Wyland, and Jennifer Bell.

Finally, thanks to the worldwide GIS user community for doing amazing work with ArcGIS technology.

Credits

Your Big Ideas: Start Here